MW00760198

NANOTECHNOLOGY SCIENCE AND TECHNOLOGY

NANOCOMMUNICATION NETWORKS

Nanotechnology Science and Technology

Additional books in this series can be found on Nova's website under the Series tab.

Additional E-books in this series can be found on Nova's website under the E-book tab.

Lasers and Electro-Optics Research and Technology

Additional books in this series can be found on Nova's website under the Series tab.

Additional E-books in this series can be found on Nova's website under the E-book tab.

NANOTECHNOLOGY SCIENCE AND TECHNOLOGY

NANOCOMMUNICATION NETWORKS

PREECHA P. YUPAPIN,
SOMSAK MITATHA,
JALIL ALI,
AND
CHAT TEEKA

Nova Science Publishers, Inc.
New York

For permission to use material from this book please contact us:
Telephone 631-231-7269; Fax 631-231-8175
Web Site: http://www.novapublishers.com

NOTICE TO THE READER

Additional color graphics may be available in the e-book version of this book.

Library of Congress Cataloging-in-Publication Data

Nanocommunication networks / editors, Preecha P. Yupapin and Somsak Mitatha.
p. cm.
Includes bibliographical references and index.
ISBN 978-1-61470-812-4 (hardcover)
1. Nanonetworks. I. Yupapin, Preecha P. II. Mitatha, Somsak.
TK7874.845.N36 2011
004.6--dc23

2011027359

Published by Nova Science Publishers, Inc. † New York

CONTENTS

PREFACE

Optical device is become the interesting tool which can be involved in various applications. The use of such device has been investigated in many areas, for instance, E/O (electrical/optical) and O/E(optical/electrical) signal converters, optical signal processing, optical sensor, optical communication and medicine, etc. has been realized. More interesting applications have been introduced, especially, the use of micro/nano scale device, in which many aspects of investigation become the challenge and interesting. In this book, we investigate the behavior of light(light pulse) within the micro and nano scale device(ring resonator), which can be integrated to form the device, circuits and systems that can be used for atom/molecule trapping and transportation, optical transistor, fast calculation device(optical gate), nanoscale communication and networks, and a device for medical applications.

P.P. Yupapin
Chat Teeka
Somsak Mitatha
Jalil Ali

Chapter 1

LIGHT PULSE IN MICRO/NANO RING RESONATOR

1.1. INTRODUCTION

A Gaussian pulse has been recognized in the form of a laser pulse that can be used in both theoretical and experimental investigation in many subjects. However, in some ways, the limit of laser power cause a problem, especially, when the high output power or long distance link is required. Optical soliton becomes a powerful tool that can overcome such a problem, i.e. for high power laser source. Furthermore, the non-dispersion of soliton in medium is the other advantage. Optical solitons can naturally be divided into classes of dark and bright solitons, whereas a dark soliton exhibits an interesting and remarkable behavior, when it is transmitted into an optical transmission system. It has the advantage of the signal detection difficulty, when the ambiguity of signal detection becomes a problem for the un-wanted users. In principle, the soliton generations and their behaviors in media are well analyzed and described by Agarwal [1]. Many earlier theoretical and experimental works on soliton applications can be found in the soliton application book by Hasegawa [2]. However, to make such a tool more useful, the problems of soliton-soliton interactions [3], collision [4], rectification [5], and dispersion management [6-8] must be solved and addressed. Therefore, in this work, we are looking for a powerful laser source with broad spectrum that can be used in many applications.

Recently, several research works have shown that use of dark and bright soliton in various applications can be realized [9-14], where one of them has shown that the secured signals in the communication link can be retrieved by using a suitable an add/drop filter that is connected into the transmission line. The other promising application of a dark soliton signal [15] is for the large guard band of two different frequencies which can be achieved by using a dark soliton generation scheme and trapping a dark soliton pulse within a nano ring resonator [1]. Furthermore, the dark soliton pulse shows a more stable behavior than the bright solitons with respect to the perturbations such as amplifier noise, fiber losses, and intra-pulse stimulated Raman scattering [16] It is found that the dark soliton pulses propagation in a lossy fiber, spreads in time at approximately half the rate of bright solitons. The dark solitons trapped in add/drop system is realized [17]. In this chapter, the use of three forms of laser pulses, i.e. Gaussian soliton, dark and bright soliton propagating within the proposed ring resonator systems is investigated and described. The use of suitable simulation parameters based on the realistic device is discussed. The potential application for new laser sources, new communication bands and dynamic optical tweezers is also discussed.

1.2. Theoretical Background

1.2.1. Gaussian Pulse

Light from a monochromatic light source is launched into a ring resonator with constant light field amplitude (E_0) and random phase modulation as shown in Figure 1.1, which is the combination of terms in attenuation (α) and phase(ϕ_0) constants, which results in temporal coherence degradation. Hence, the time dependent input light field (E_{in}), without pumping term, can be expressed as [18]

$$E_{in}(t) = E_0 e^{-\alpha L + j\phi_0(t)} \tag{1.1}$$

where L is a propagation distance(waveguide length).

We assume that the nonlinearity of the optical ring resonator is of the Kerr-type, i.e., the refractive index is given by

$$n = n_0 + n_2 I = n_0 + \left(\frac{n_2}{A_{eff}}\right)P, \tag{1.2}$$

where n_0 and n_2 are the linear and nonlinear refractive indexes, respectively. I and P are the optical intensity and optical power, respectively. The effective mode core area of the device is given by A_{eff}. For the microring and nanoring resonators, the effective mode core areas range from 0.10 to 0.50 μm^2 [19, 20].

When a Gaussian pulse is input and propagated within a fiber ring resonator, the resonant output is formed, thus, the normalized output of the light field is the ratio between the output and input fields ($E_{out}(t)$ and $E_{in}(t)$) in each roundtrip, which can be expressed as [21]

$$\left|\frac{E_{out}(t)}{E_{in}(t)}\right|^2 = (1-\gamma)\left[1-\frac{(1-(1-\gamma)x^2)\kappa}{(1-x\sqrt{1-\gamma}\sqrt{1-\kappa})^2+4x\sqrt{1-\gamma}\sqrt{1-\kappa}\sin^2(\frac{\phi}{2})}\right] \tag{1.3}$$

Equation (1.3) indicates that a ring resonator in the particular case is very similar to a Fabry-Perot cavity, which has an input and output mirror with a field reflectivity, $(1-\kappa)$, and a fully reflecting mirror. k is the coupling coefficient, κ and $x=\exp(-\alpha L/2)$ represents a roundtrip loss coefficient, $\phi_0 = kLn_0$ and $\phi_{NL} = kL(\frac{n_2}{A_{eff}})P$ are the linear and nonlinear phase shifts,

$k = 2\pi/\lambda$ is the wave propagation number in a vacuum. Where L and α are a waveguide length and linear absorption coefficient, respectively. In this work, the iterative method is introduced to obtain the results as shown in equation (1.3), similarly, when the output field is connected and input into the other ring resonators.

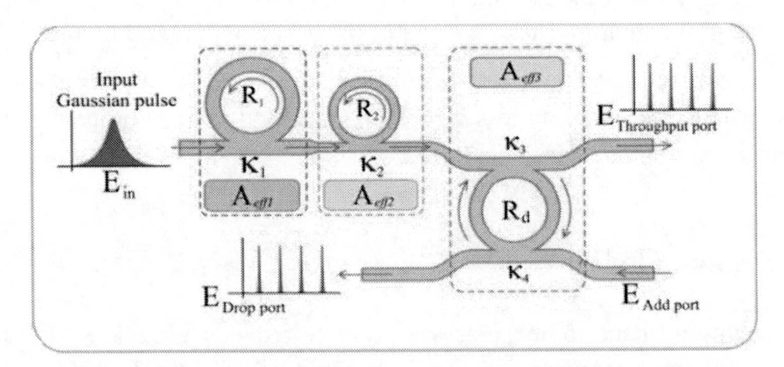

Figure 1.1. A schematic of a Gaussian soliton generation system, where R_s: ring radii, κ_s: coupling coefficients, R_d: an add/drop ring radius, A_{effs}: Effective areas.

The input optical field as shown in equation (1.1), i.e. a Gaussian pulse, is input into a nonlinear microring resonator. By using the appropriate parameters, the chaotic signal is obtained by using equation (1.3). To retrieve the signals from the chaotic noise, we propose to use the add/drop device with the appropriate parameters. This is given in details as followings. The optical outputs of a ring resonator add/drop filter can be given by the equations (1.4) and (1.5).

$$\left|\frac{E_t}{E_{in}}\right|^2 = \frac{(1-\kappa_1) - 2\sqrt{1-\kappa_1}\cdot\sqrt{1-\kappa_2}\,e^{-\frac{\alpha}{2}L}\cos(k_n L) + (1-\kappa_2)e^{-\alpha L}}{1+(1-\kappa_1)(1-\kappa_2)e^{-\alpha L} - 2\sqrt{1-\kappa_1}\cdot\sqrt{1-\kappa_2}\,e^{-\frac{\alpha}{2}L}\cos(k_n L)} \tag{1.4}$$

and

$$\left|\frac{E_d}{E_{in}}\right|^2 = \frac{\kappa_1\kappa_2 e^{-\frac{\alpha}{2}L}}{1+(1-\kappa_1)(1-\kappa_2)e^{-\alpha L} - 2\sqrt{1-\kappa_1}\cdot\sqrt{1-\kappa_2}\,e^{-\frac{\alpha}{2}L}\cos(k_n L)} \tag{1.5}$$

where E_t and E_d represents the optical fields of the throughput and drop ports respectively. The transmitted output can be controlled and obtained by choosing the suitable coupling ratio of the ring resonator, which is well derived and described by reference [21]. Where $\beta = kn_{eff}$ represents the propagation constant, n_{eff} is the effective refractive index of the waveguide, and the circumference of the ring is $L = 2\pi R$, here R is the radius of the ring. The chaotic noise cancellation can be managed by using the specific parameters of the add/drop device, which the required signals at the specific wavelength band can be filtered and retrieved. $K_1$and K_2 are coupling coefficient of add/drop filters, $k_n = 2\pi/\lambda$ is the wave propagation number for in a vacuum, and the waveguide (ring resonator) loss is $\alpha = 0.5$ dBmm^{-1}. The fractional coupler intensity loss is $\gamma = 0.1$. In the case of add/drop device, the nonlinear refractive index is neglected.

1.2.2. Soliton Pulse

Bright and dark soliton pulses are introduced into the multi-stage nano ring resonators as shown in Figure 1.2, the input optical field (E_{in}) of the bright and dark soliton pulses input are given by an Eq. (1.6) and (1.7)[1], respectively.

$$E_{in}(t) = A\sec h\left[\frac{T}{T_0}\right]\exp\left[\left(\frac{z}{2L_D}\right) - i\omega_0 t\right] \tag{1.6}$$

and

$$E_{in}(t) = A\tanh\left[\frac{T}{T_0}\right]\exp\left[\left(\frac{z}{2L_D}\right) - i\omega_0 t\right] \tag{1.7}$$

Here A and z are the optical field amplitude and propagation distance, respectively. T is a soliton pulse propagation time in a frame moving at the group velocity, $T=t-\beta_1 z$, where β_1 and β_2 are the coefficients of the linear and second-order terms of Taylor expansion of the propagation constant. $L_D = T_0^2/|\beta_2|$ is the dispersion length of the soliton pulse. T_0 in equation is a soliton pulse propagation time at initial input (or soliton pulse width), where t is the soliton phase shift time, and the frequency shift of the soliton is ω_0. This solution describes a pulse that keeps its temporal width invariance as it propagates, and thus is called a temporal soliton. When a soliton peak intensity $(|\beta_2/\Gamma T_0^2|)$ is given, then T_0 is known. For the soliton pulse in the microring device, a balance should be achieved between the dispersion length (L_D) and the nonlinear length ($L_{NL}=1/\Gamma\phi_{NL}$), where $\Gamma=n_2 k_0$, is the length scale over which dispersive or nonlinear effects makes the beam become wider or narrower.

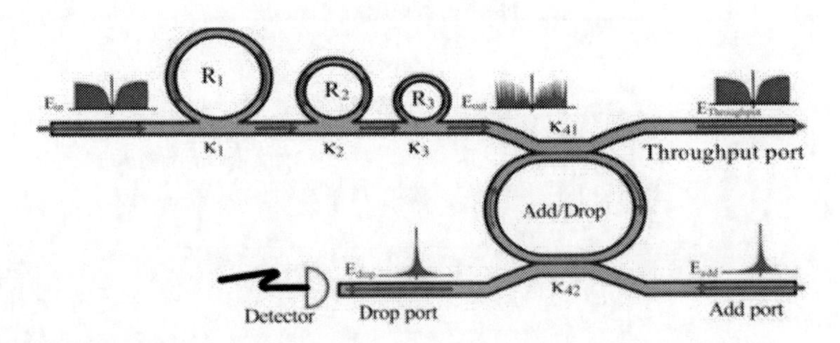

Figure 1.2. Schematic of a dark-bright soliton conversion system, where R_s is the ring radii, κ_s is the coupling coefficient, and κ_{41} and κ_{42} are the add/drop coupling coefficients.

For a soliton pulse, there is a balance between dispersion and nonlinear lengths, hence $L_D = L_{NL}$. Similarly, the output soliton of the system in Figure 2 can be calculated by using Gaussian equations as given in the above case.

1.3. SIMULATION RESULTS

1.3.1. Gaussian Pulse

From Figure 1.1, in principle, light pulse is sliced to be the discrete signal and amplified within the first ring, where more signal amplification can be obtained by using the smaller ring device (second ring).

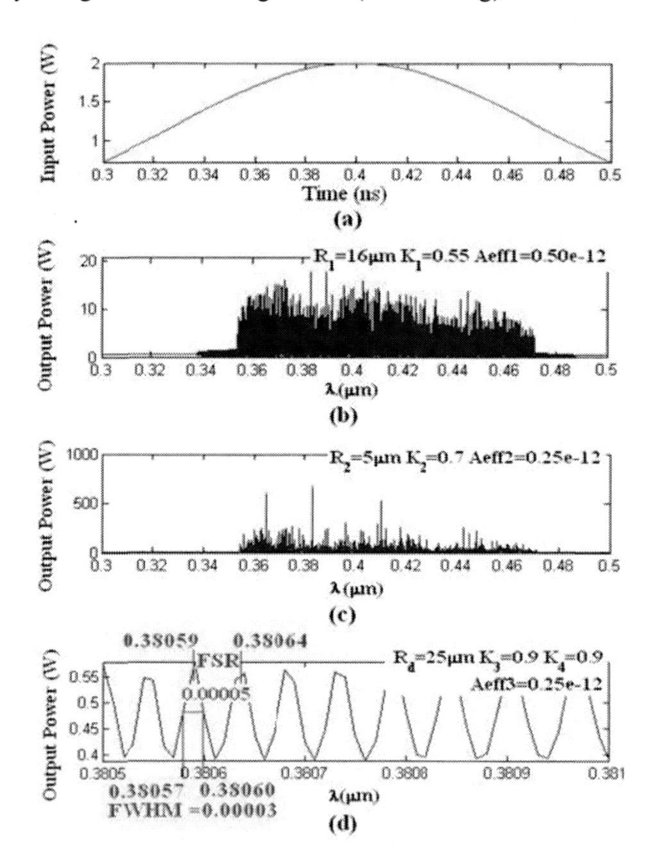

Figure 1.3. Result of the spatial pulses with center wavelength at 0.40 μm, where (a) the Gaussian pulse, (b) large bandwidth signals, (c) large amplified signals, (d) filtering and amplifying signals from the drop port.

Finally, the required signals can be obtained via a drop port of the add/drop filter. In operation, an optical field in the form of Gaussian pulse from a laser source at the specified center wavelength is input into the system. From Figure 1.3, the Gaussian pulse with center wavelength (λ_0) at 0.40 μm, pulse width (Full Width at Half Maximum, FWHM) of 20 ns, peak power at 2 W is input into the system as shown in Figure 1.3(a).

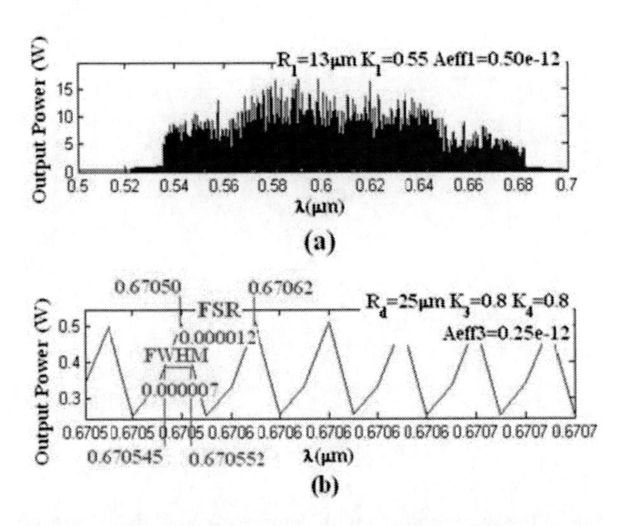

Figure 1.4. Result of the spatial pulses with center wavelength at 0.60 μm, where (a) large bandwidth signals, (b) filtering and amplifying signals from the drop port.

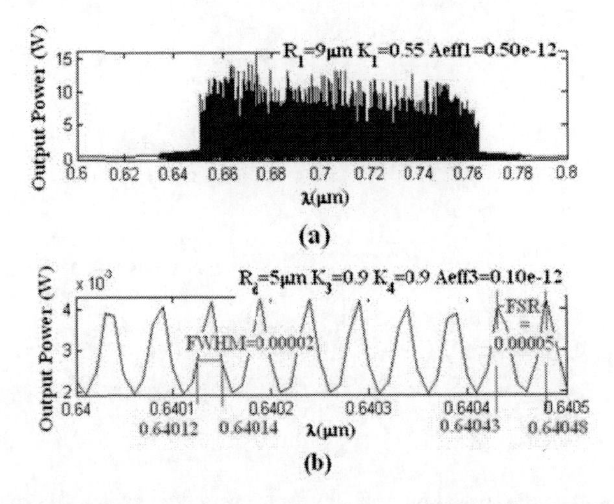

Figure 1.5. Result of the spatial pulses with center wavelength at 0.70 μm, where (a) large bandwidth signals, (b) filtering and amplifying signals from the drop port.

The large bandwidth signals can be seen within the first microring device, and shown in Figure 1.3(b). The suitable ring parameters are used, for instance, ring radii R_1= 16.0 μm, R_2= 5.0 μm, and R_d= 25.0 μm. In order to make the system associate with the practical device [19, 20], the selected parameters of the system are fixed to n_0 = 3.34 (InGaAsP/InP), A_{eff} = 0.50 μm^2 and 0.25 μm^2 for a microring and add/drop ring resonator, respectively, α = 0.5 $dBmm^{-1}$, γ = 0.1.

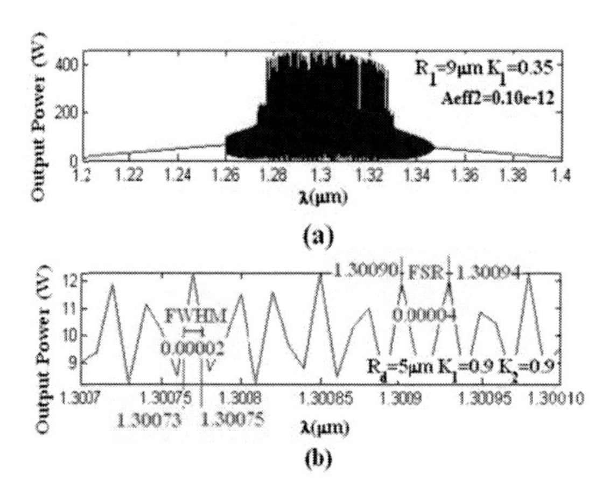

Figure 1.6. Result of the spatial pulses with center wavelength at 1.30 μm, where (a) large bandwidth signals, (b) filtering and amplifying signals from the drop port.

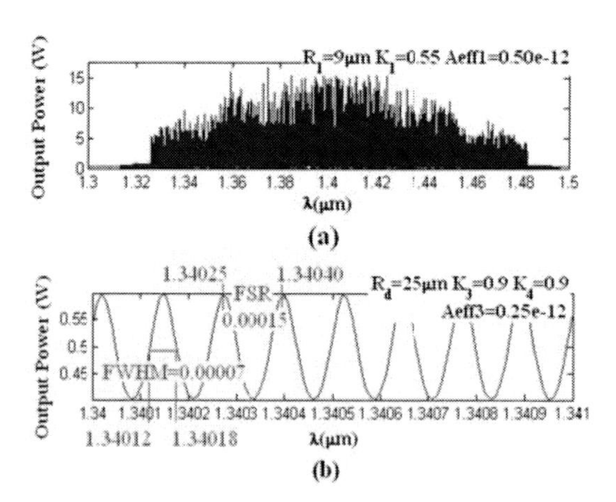

Figure 1.7. Result of the spatial pulses with center wavelength at 1.40 μm, where (a) large bandwidth signals, (b) filtering and amplifying signals from the drop port.

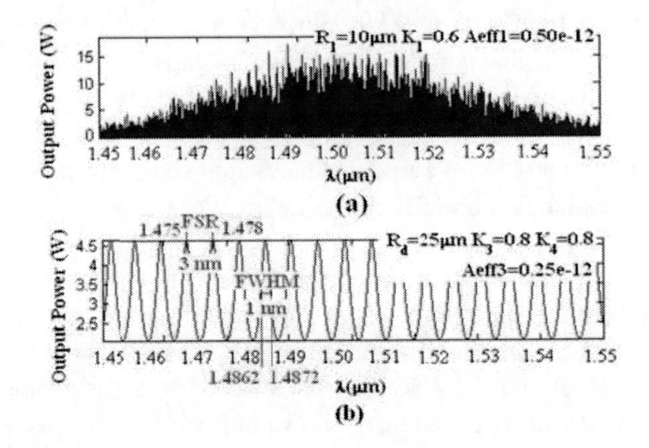

Figure 1.8. Results of the spatial pulses with center wavelength at 1.50 µm, where (a) large bandwidth signals, (b) filtering and amplifying signals from the drop port.

In this investigation, the coupling coefficient (kappa, κ) of the microring resonator is ranged from 0.55 to 0.90. The nonlinear refractive index of the microring used is n_2=2.2 x 10^{-17} m^2/W. In this case, the attenuation of light propagates within the system (i.e. wave guided) used is 0.5dBmm^{-1}. After light is input into the system, the Gaussian pulse is chopped (sliced) into a smaller signal spreading over the spectrum due to the nonlinear effects[22], which is shown in Figure 1.3(a). The large bandwidth signal is generated within the first ring device. In applications, the specific input or out wavelengths can be used and generated. For instance, the different center wavelengths of the input pulse can be ranged from 0.40-1.5 µm as shown in Figs. 1.3-1.8, where the suitable parameters are used and shown in the figures.

1.3.2. Dark Soliton

In operation, a dark soliton pulse with 50-ns pulse width with the maximum power of 0.65W is input into the dark-bright soliton conversion system as shown in Figure 1.2. The suitable ring parameters are ring radii, where R_1=10.0µm, R_2=7.0µm, and R_3=5.0µm. In order to make the system associate with the practical device [19, 20] the selected parameters of the system are fixed to λ_0=1.50µm, n_0=3.34 (InGaAsP/InP). The effective core areas are A_{eff}=0.50, 0.25, and 0.10 µm^2 for a microring resonator (MRR) and nanoring resonator (NRR), respectively. The waveguide and coupling loses are α =0.5dBmm^{-1} and γ =0.1, respectively, and the coupling coefficients κ_s of the

MRR are ranged from 0.05 to 0.90. However, more parameters are used as shown in Figure 1.2. The nonlinear refractive index is n_2=2.2×10^{-13} m^2/W. In this case, the waveguide loss used is 0.5 dBmm^{-1}. The input dark soliton pulse is chopped (sliced) into the smaller output signals of the filtering signals within the rings R_2 and R_3. We find that the output signals from R_3 are smaller than from R_1, which is more difficult to detect when it is used in the link. In fact, the multistage ring system is proposed due to the different core effective areas of the rings in the system, where the effective areas can be transferred from 0.50 to 0.10μm^2 with some losses. The soliton signals in R_3 is entered in the add/drop filter, where the dark-bright soliton conversion can be performed by using Eqs. (1.6) and (1.7). Results obtained when a dark soliton pulse is input into a MRR and NRR system as shown in Figure 1.9. The add/drop filter is formed by two couplers and a ring radius (R_d) of 10μm, the coupling constants (κ_{41} and κ_{42}) are the same values (0.50). When the add/drop filter is connected to the third ring (R_3), the dark-bright soliton conversion can be seen. The bright and dark solitons are detected by the through (throughput) and drop ports as shown in Figure 1.9(d)-(e), respectively.

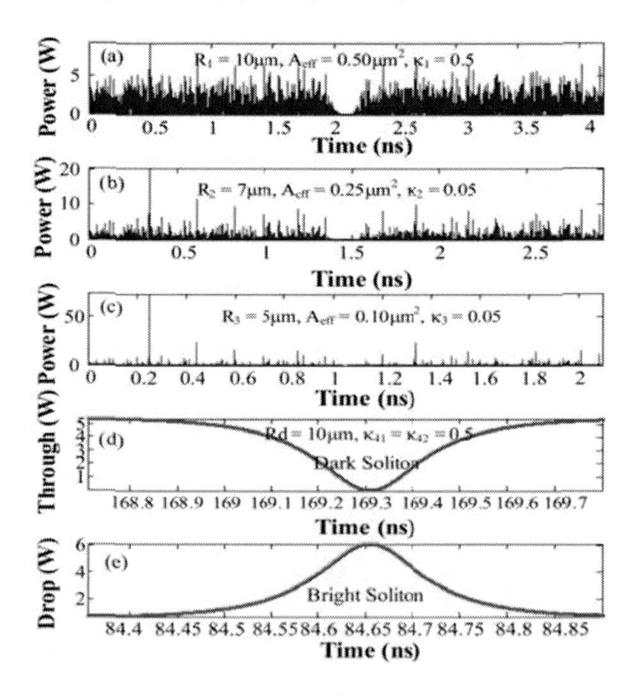

Figure 1.9. Results of the soliton signals within the ring resonator system, where (a) in ring R_1, (b) in ring R_2, (c) – (d) in ring R_3, and (d) – (e) dark – bright solitons conversion at the add/drop filter. The input dark soliton power is 2W.

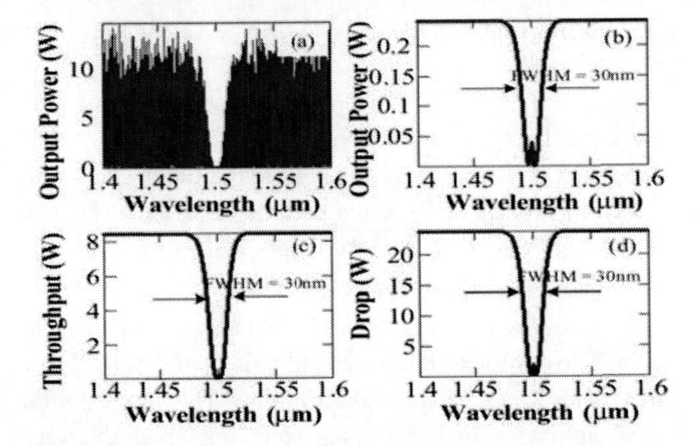

Figure 1.10. Shows the dynamic dark soliton(optical tweezers) within the add/drop filter, when the bright soliton is input into the add port with the center wavelength λ_0 = 1.5μm. (a) add/drop signals, (b) dark – bright soliton collision, (c) optical tweezers at throughput port, and (d) optical tweezers at drop port.

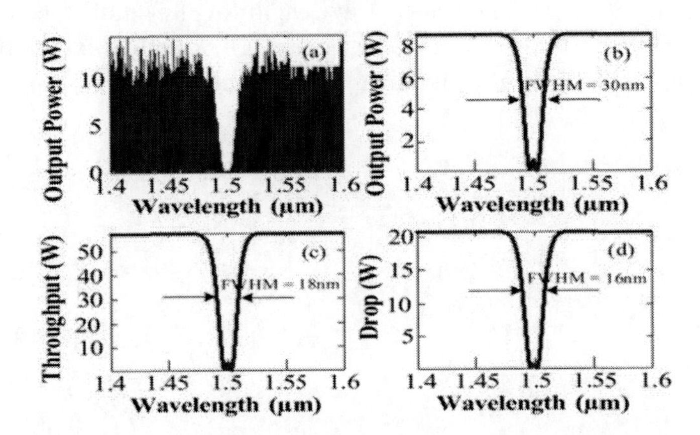

Figure 1.11. The dynamic dark soliton(optical tweezers) occurs within add/drop tunable filter, when the bright soliton is input into the add port with the center wavelength λ_0 = 1.5μm. (a) add/drop signals, (b) dark – bright soliton collision, (c) optical tweezers at throughput port, and (d) optical tweezers at drop port.

1.3.3. Bright Soliton

The large bandwidth signal within the micro ring device can be generated by using a common soliton pulse input into the nonlinear micro ring resonator. This means that the broad spectrum of light can be generated after the soliton pulse is input into the ring resonator system. The schematic diagram of the proposed system is as shown in Figure 1.12. A soliton pulse with 50 ns pulse width, peak power at 2 W is input into the system. The suitable ring parameters are used, for instance, ring radii R_1= 15.0 μm, R_2= 10.0 μm, R_3= R_s=5.0 μm and R_5=R_d= 20.0 μm. In order to make the system associate with the practical device [19, 20], the selected parameters of the system are fixed to λ_0 = 1.55 μm, n_0 = 3.34 (InGaAsP/InP), A_{eff} = 0.50, 0.25 μm^2 and 0.10 μm^2 for a micro ring and nano ring resonator, respectively, α = 0.5 $dBmm^{-1}$, γ = 0.1. The coupling coefficient (kappa, κ) of the micro ring resonator ranged from 0.1 to 0.95. The nonlinear refractive index is n_2=2.2 x 10^{-13} m^2/W. In this case, the wave guided loss used is 0.5$dBmm^{-1}$. The input soliton pulse is chopped (sliced) into the smaller signals spreading over the spectrum (i.e. broad wavelength) as shown in Figure 1.13(b) and 1.13(g), which is shown that the large bandwidth signal is generated within the first ring device. The biggest output amplification is obtained within the nano-waveguides (rings R_3 and R_4) as shown in Figs. 1.13(d) and 1.13(e), whereas the maximum power of 10 W is obtained at the center wavelength of 1.5 μm. The coupling coefficients are given as shown in the figures. The coupling loss is included due to the different core effective areas between micro and nano ring devices, which is given by 0.1dB.

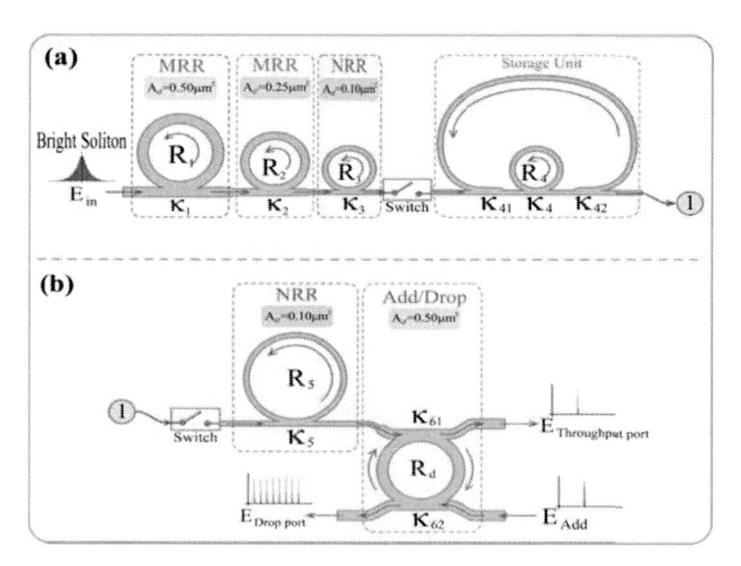

Figure 1.12. A broadband generation system, (a) a broadband source generation and a storage unit, (b) a soliton band selector, where R_s: ring radii, κ_s: coupling coefficients, κ_{41}, κ_{42}: coupling losses, k_{61} and k_{61} are the add/drop coupling coefficients.

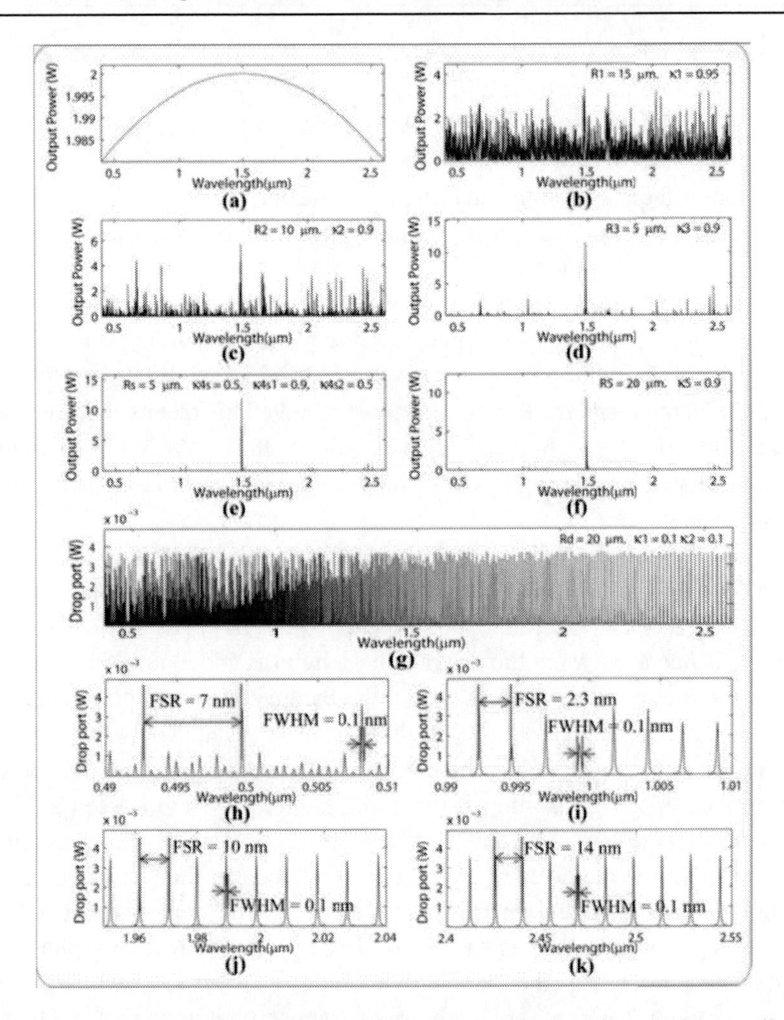

Figure 1.13. A soliton band with center wavelength at 1.5 μm, where (a) input soliton, (b) ring R_1, (c) ring R_2, (d) ring R_3, (e) storage ring(R_s), (f) ring R_5, (g) drop port output signals. The output of different soliton bands (center wavelength) are as shown, where (h) 0.51 μm, (i) 0.98 μm, (j) 1.99 μm and (k) 2.48 μm.

1.4. APPLICATIONS

The application of this proposed work can be categorized into three cases, firstly, we have shown that the multi-wavelength bands can be generated by using a Gaussian pulse propagating within the microring resonator system, which is available for the extended DWDM with the wavelength center at 400-

1,400 nm, which can be used in the existed public network. Results obtained have shown that the spatial pulses width of 30 nm and the spectrum range of 400 nm can be generated and achieved. Moreover, the problem of signal collision can be solved by using the suitable FSR design [21], while the dispersion effect is minimized when the center wavelength at 1.30 μm. Furthermore, the Gaussian pulse output power can be amplified, which can provide the power budget for long distance link. We find that the maximum power of 400 W can be obtained, however, the coupling coefficient of the add/drop filter is the major parameter of the required coupling output power, for instance, the output power of 12 W is obtained as shown in Figure 1.6(b), where the parameters are R_d= 5.0 μm, κ_3=κ_4=0.9. This means that the use of wavelength 0.4-1.40 μm (Gaussian pulse) for DWDM via optical communication is plausible, which can be used with the existed public network installation.

Secondly, the dynamic dark soliton control can be configured to be an optical dynamic tool known as an optical tweezers, where more details of optical tweezers can be found in references [23, 24]. The optical tweezers behavior is occurred when the bright soliton input is added into the system via add port as shown in Figure 1.2, where the parameters of system are used the same as the previous case. The bright soliton is generated with the central wavelength $\lambda_0 = 1.5\mu m$, when the bright soliton propagating into the add/drop system, the dark-bright soliton collision in add/drop system is seen as shown in Figure 1.10 (a)-(b). The optical tweezers probe can be trapped/confined atom/light within the well of the probe. The dark soliton valley dept, i.e. potential well is changed when it is modulated by the trapping energy (dark-bright solitons interaction) as shown in Figure 1.11. The recovery photon can be obtained by using the dark-bright soliton conversion, which is well analyzed by Sarapat et al. [14], where the trapped photon or molecule can be released or separated from the dark soliton pulse, in practice, in this case the bright soliton is become alive and seen.

Finally, we have shown that a large bandwidth of the optical signals with the specific wavelength can be generated within the micro ring resonator system as shown in Figure 1.12. The amplified signals with broad spectrum can be generated, stored and regenerated within the nano-waveguide. The maximum stored power of 10 W is obtained as shown in Figure 1.13(d) and 1.13(e), where the average regenerated optical output power of 4 W is achieved via and a drop port of an add/drop filter as shown in Figure 1.13(h)-1.13(k), which is a broad spectra of light cover the large bandwidth as shown in Figure 1.13(g). However, to make the system being realistic, the waveguide and connection losses are required to address in the practical device, which may be affected the signal amplification. The storage light pulse within a storage ring (R_s or R_4) is achieved, which has also been reported by Ref. [11]. In applications, the increasing in communication channel and network capacity can be formed by using the different soliton bands (center wavelength) as shown in Figure 1.13, where 1.13(h) 0.51 μm, 1.13(i) 0.98 μm , 1.13(j) 1.48 μm and 1.13(k) 2.46 μm are the generated center wavelengths of the soliton bands. The selected wavelength center can be performed by using the designed add/drop filter, where the required spectral width (Full Width at Half Maximum, FWHM) and free spectrum range (FSR) are obtained, the channel spacing and bandwidth are represented by FSR and FWHM, respectively, for instance, the FSR and FWHM of 2.3 nm and 100 pm are obtained as shown in Figure 1.13(i).

1.5. Conclusion

We have demonstrated that some interesting results can be obtained when the laser pulse is propagated within the nonlinear optical ring resonator, especially, in microring and nanoring resonators, which can be used to perform many applications. For instance, the broad spectrum of a monochromatic source with the reasonable power can be generated and achieved by using a Gaussian pulse, where a dark soliton can be converted to be a bright soliton by using the ring resonator system incorporating the add/drop multiplexer, which can be cofigured as a dynamic optical tweezers. Moreover, the use of a bright soliton can provide the non-dispersion soliton, where the generation of soliton communication bandwidth with the center wavelength at 1.30 μm is achieved.

References

[1] G. P. Agarwal, *Nonlinear Fiber Optics*, Academic Press, 4th edition, New York, 2007.

[2] A. Hasegawa, Ed., Massive, *WDM and TDM Soliton Transmission Systems,* Kluwer Academic Publishers, 2000.

[3] Yu. A. Simonov, and J. A. Tjon, "Soliton-soliton interaction in confining models," *Phys. Lett. B*, 85, 380–384(1979).

[4] J. K. Drohm, L. P. Kok, Yu. A. Simonov, J. A. Tjon, and A. I. Veselov, "Collision, and rotation of solitons in three space-time dimensions," *Phys. Lett. B*, 101, 204–208(1981).

[5] T. Iizuka and Yu. S. Kivshar, "Optical gap solitons in nonresonant quadratic media," *Phys. Rev. E,* 59, 7148–7151(1999).

[6] A. Biswas, "Dispersion-managed solitons in optical fibers," *"J. of Optics A*, 4(1), 84-97. (2002).

[7] R. Kohl, D. Milovic, E. Zerrad and A. Biswas, "Soliton perturbation theory for dispersion-managed optical fibers, *J. of Nonlinear Optical Physics and Materials*," 18(2), 227-270 (2009).

[8] R. Ganapathy, K. Porsezian, A. Hasegawa, and V. N. Serkin, "Soliton interaction under soliton dispersion management," *IEEE J. Quantum Electron.* 44, 383–390(2008).

[9] N. Pornsuwancharoen, U. Dunmeekaew and P.P. Yupapin, "Multi-soliton generation using a micro ring resonator system for DWDM based soliton communication," *Microw. and Opt. Technol. Lett.*, 51(5), 1374-1377(2009).

[10] P.P. Yupapin, N. Pornsuwanchroen and S. Chaiyasoonthorn, "Attosecond pulse generation using nonlinear micro ring resonators," *Microw. and Opt. Technol. Lett.*, 50(12), 3108-3011(2008).

[11] N. Pornsuwancharoen and P.P. Yupapin, "Generalized fast, slow, stop, and store light optically within a nano ring resonator," *Microw. and Opt. Technol. Lett.*, 51(4), 899-902(2009).

[12] N. Pornsuwancharoen, S. Chaiyasoonthorn and P.P. Yupapin, "Fast and slow lights generation using chaotic signals in the nonlinear micro ring resonators for communication security," *Opt. Eng.,* 48(1), 50005-1-5(2009).

[13] [P.P. Yupapin and N. Pornsuwancharoen, "Proposed nonlinear micro ring resonator arrangement for stopping and storing light," *IEEE Photon. Technol. Lett.,* 21, 404-406(2009).

[14] K. Sarapat, N. Sangwara, K. Srinuanjan, P.P. Yupapin and N. Pornsuwancharoen, "Novel dark-bright optical solitons conversion system and power amplification," *Opt. Eng.*, 48, 045004(1-7) (2009).

[15] S. Mitatha, N. Pornsuwancharoen and P.P.Yupapin, "A simultaneous short-wave and millimeter-wave generation using a soliton pulse within a nano-waveguide*", IEEE Photon. Technol. Lett.*, 21, 932-934 (2009).

[16] M.E. Heidari, M.K. Moravvej-Farshi, and A. Zariffkar, "Multichannel wavelength conversion using fourth-order soliton decay", *J. Lightwave Technol.,* 25, 2571-2578 (2007).

[17] A. Charoenmee, N. Pornsuwancharoen and P.P.Yupapin, "Trapping a dark soliton pulse within a nano ring resonator", *International J. of Light and Electron Optics,* (2009). doi:10.1016/j.ijleo.2009.03.015.

[18] D. Deng and Q. Guo, "Ince-Gaussian solitons in strongly nonlocal nonlinear media", *Opt. Lett.,* 32, 3206-3208(2007).

[19] Y. Kokubun, Y. Hatakeyama, M. Ogata, S. Suzuki, and N. Zaizen, "Fabrication technologies for vertically coupled micro ring resonator with multilevel crossing busline and ultracompact-ring radius," *IEEE J. Sel. Top. Quantum Electron.* 11, 4–10(2005).

[20] Y. Su, F. Liu, and Q. Li, "System performance of slow-light buffering, and storage in silicon nano-waveguide," *Proc. SPIE* 6783, 67832P(2007).

[21] P.P. Yupapin, P. Saeung and C. Li, "Characteristics of complementary ring-resonator add/drop filters modeling by using graphical approach," *Opt. Commun.*, 272, 81-86(2007).

[22] P.P. Yupapin and W. Suwancharoen, "Chaotic signal generation and cancellation using a micro ring resonator incorporating an optical add/drop multiplexer," *Opt. Commun.,* 280/2, 343-350(2007).

[23] L. Yuan, Z. Liu, J. Yang and C. Guan, "Twin-core fiber optical tweezers", *Opt. Exp.*, 16, 4559-4566 (2008).

[24] N. Malagninoa, G. Pescea, A. Sassoa and E. Arimondo, "Measurements of trapping efficiency and stiffness in optical tweezers", *Opt. Commun.,* 214, 15-24 (2002).

Chapter 2

PANDA RING RESONATOR

2.1. INTRODUCTION

The use of Gaussian pulse has reported the interesting results of light pulse propagating within a nonlinear media [1, 2]. In the latter work, they have shown that the transfer function of the output at resonant condition is derived and studied. They found that the broad spectrum of light pulse can be transformed to the discrete pulses. In this work we are looking for a common laser source that can be used to extend the used/installed wavelength band in the public network, especially, when with the broad center wavelengths can be generated and enhanced. Therefore, the use of a Gaussian pulse to form a broad wavelength bands within a microring resonator system is recommended, whereas the other problems are such as output power and signal collision that can be solved [3, 4, 5]. One of the obtained results has shown that by using the center wavelength at 1,550 nm with suitable parameters, the broad Gaussian pulses generation is plausible. Moreover, most of the results have shown that the optical signals, i.e. Gaussian pulse can be amplified and enhanced within the nonlinear ring resonator system, which is known as a double PANDA ring resonator system. The dynamic propagation of pulses within the desired system is also shown and discussed in details.

2.2. A PANDA RING RESONATOR

Light from a monochromatic light source is launched into a ring resonator with constant light field amplitude (E_0) and random phase modulation as shown in Figure 2.1, which is the combination of terms in attenuation (α) and phase (φ_0) constants, which results in temporal coherence degradation. Hence, the time dependent input light field (E_{in}), without pumping term, can be expressed as [6]

$$E_{in}(t) = E_0 \exp\left[-\alpha L + j\varphi_0(t)\right] \quad (2.1)$$

where L is a propagation distance(waveguide length).

When light propagates within the nonlinear material (medium), the refractive index (n) of light within the medium is given by

$$n = n_0 + n_2 I = n_0 + \frac{n_2}{A_{eff}} P, \quad (2.2)$$

where n_0 and n_2 are the linear and nonlinear refractive indexes, respectively. I and P are the optical intensity and optical power, respectively. The effective mode core area of the device is given by A_{eff}.

In Figure 2.1, consists of add/drop optical multiplexing used for generated random binary coded light pulse and other is add/drop optical filter device for decoded binary code signal. The resonator output field, E_{t1} and E_1 consists of the transmitted and circulated components within the add/drop optical multiplexing system, which can perform the driven force to photon/molecule/atom.

(a)

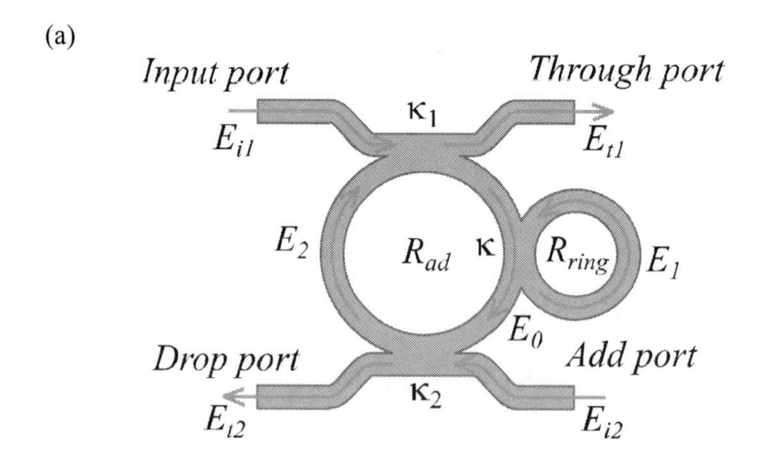

(b)

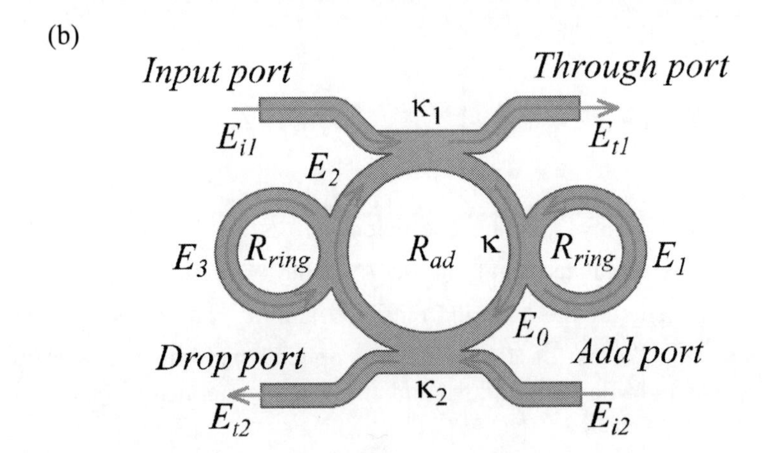

Figure 2.1. A schematic of ring resonator system, where (a) Semi-PANDA ring resonator, (b) PANDA ring resonators.

When the input light pulse passes through the first coupling device of the add/drop optical multiplexing system, the transmitted and circulated components can be written as

$$E_{t1} = \sqrt{1-\gamma_1}\left[\sqrt{1-\kappa_1}E_{i1} + j\sqrt{\kappa_1}E_4\right] \tag{2.3}$$

$$E_1 = \sqrt{1-\gamma_1}\left[\sqrt{1-\kappa_1}E_4 + j\sqrt{\kappa_1}E_{i1}\right] \tag{2.4}$$

$$E_2 = E_{R2}e^{-\frac{\alpha}{2}\frac{L}{2} - jk_n\frac{L}{2}} \tag{2.5}$$

where κ_1 and γ_1 are the intensity coupling coefficient and the fractional coupler intensity loss of the add/drop optical filter, respectively. α is the attenuation coefficient, $k_n = 2\pi/\lambda$ is the wave propagation number, λ is the input wavelength light field and $L = 2\pi R_{ad}$, R_{ad} is the radius of add/drop device.

For the second coupler of the add/drop optical multiplexing system,

$$E_{t2} = \sqrt{1-\gamma_2}\left[\sqrt{1-\kappa_2}E_{i2} + j\sqrt{\kappa_2}E_2\right] \tag{2.6}$$

$$E_3 = \sqrt{1-\gamma_2}\left[\sqrt{1-\kappa_2}E_2 + j\sqrt{\kappa_2}E_{i2}\right] \tag{2.7}$$

$$E_4 = E_{R1}e^{-\frac{\alpha}{2}\frac{L}{2}-jk_n\frac{L}{2}} \tag{2.8}$$

where κ_2 is the intensity coupling coefficient, γ_2 is the fractional coupler intensity loss. The circulated light fields, E_{R1} and E_{R2} are the light field circulated components of the nanoring radii, R_1 and R_2 which coupled into the left and right sides of the add/drop optical multiplexing system, respectively. The light field transmitted and circulated components in the right nanoring, R_2, are given by

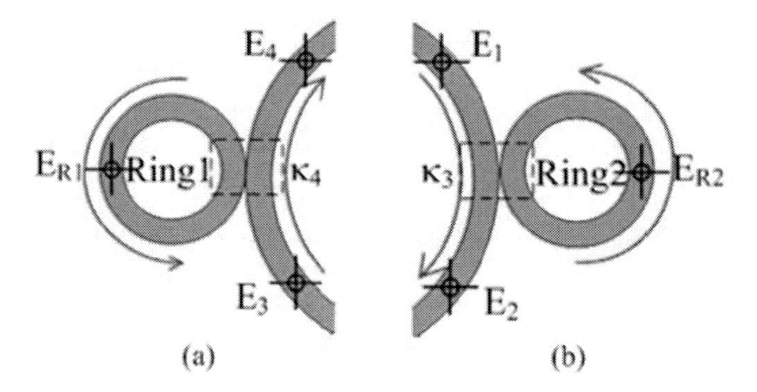

Figure 2.2. A schematic of PANDA ring, where (a) the left nanoring and (b) the right nanoring.

$$E_{R2} = \sqrt{1-\gamma_3}\left[\sqrt{1-\kappa_3}E_1 + j\sqrt{\kappa_3}E_{r2}\right] \tag{2.9}$$

$$E_{r1} = \sqrt{1-\gamma_3}\left[\sqrt{1-\kappa_3}E_{r2} + j\sqrt{\kappa_3}E_1\right] \tag{2.10}$$

$$E_{r2} = E_{r1}e^{-\frac{\alpha}{2}L_2-jk_nL_2} \tag{2.11}$$

where κ_3 and γ_3 are the intensity coupling coefficient and the fractional coupler intensity loss of the right nanoring, respectively. α is the attenuation coefficient, $k_n = 2\pi/\lambda$ is the wave propagation number, λ is the input wavelength light field and $L_2 = 2\pi R_2$, R_2 is the radius of right nanoring.

From equations (2.9)-(2.11), the circulated roundtrip light fields of the right nanoring radii, R_2, are given in equations (2.12) and (2.13), respectively.

$$E_{r1} = \frac{j\sqrt{1-\gamma_3}\sqrt{\kappa_3}E_1}{1-\sqrt{1-\gamma_3}\sqrt{1-\kappa_3}e^{-\frac{\alpha}{2}L_2-jk_nL_2}} \tag{2.12}$$

$$E_{r2} = \frac{j\sqrt{1-\gamma_3}\sqrt{\kappa_3}E_1e^{-\frac{\alpha}{2}L_2-jk_nL_2}}{1-\sqrt{1-\gamma_3}\sqrt{1-\kappa_3}e^{-\frac{\alpha}{2}L_2-jk_nL_2}} \tag{2.13}$$

Thus, the output circulated light field, E_{R2}, for the right nanoring is given by

$$E_{R2} = E_1\left\{\frac{\sqrt{(1-\gamma_3)(1-\kappa_3)}-(1-\gamma_3)e^{-\frac{\alpha}{2}L_2-jk_nL_2}}{1-\sqrt{(1-\gamma_3)(1-\kappa_3)}e^{-\frac{\alpha}{2}L_2-jk_nL_2}}\right\}. \tag{2.14}$$

Similarly, the output circulated light field, E_{R1}, for the left nanoring at the left side of the add/drop optical multiplexing system is given by

$$E_{R1} = E_3\left\{\frac{\sqrt{(1-\gamma_4)(1-\kappa_4)}-(1-\gamma_4)e^{-\frac{\alpha}{2}L_1-jk_nL_1}}{1-\sqrt{(1-\gamma_4)(1-\kappa_4)}e^{-\frac{\alpha}{2}L_1-jk_nL_1}}\right\}. \tag{2.15}$$

where κ_4 is the intensity coupling coefficient, γ_4 is the fractional coupler intensity loss, α is the attenuation coefficient, $k_n = 2\pi/\lambda$ is the wave propagation number, λ is the input wavelength light field and $L_1 = 2\pi R_1$, R_1 is the radius of left nanoring.

From equations (2.3)-(2.15), the circulated light fields, E_1, E_3 and E_4 are defined by given $x_1 = (1-\gamma_1)^{1/2}$, $x_2 = (1-\gamma_2)^{1/2}$, $y_1 = (1-\kappa_1)^{1/2}$, and $y_2 = (1-\kappa_2)^{1/2}$.

$$E_1 = \frac{jx_1\sqrt{\kappa_1}E_{i1} + jx_1x_2y_1\sqrt{\kappa_2}E_{R1}E_{i2}e^{-\frac{\alpha}{4}\frac{L}{4}-jk_n\frac{L}{4}}}{1 - x_1x_2y_1y_2E_{R2}E_{R1}e^{-\frac{\alpha}{2}L-jk_nL}} \tag{2.16}$$

$$E_3 = x_2y_2E_{R2}E_1e^{-\frac{\alpha}{4}\frac{L}{2}-jk_n\frac{L}{2}} + jx_2\sqrt{\kappa_2}E_{i2} \tag{2.17}$$

$$E_4 = x_2y_2E_{R2}E_{R1}E_1e^{-\frac{\alpha}{2}L-jk_nL} + jx_2\sqrt{\kappa_2}E_{R1}E_{i2}e^{-\frac{\alpha}{4}\frac{L}{4}-jk_n\frac{L}{4}} \tag{2.18}$$

Thus, from equations (2.3), (2.5), (2.16)-(2.18), the output optical field of the through port (E_{t1}) is expressed by

$$E_{t1} = x_1y_1E_{i1} + \left(jx_1x_2y_2\sqrt{\kappa_1}E_{R2}E_{R1}E_1 - x_1x_2\sqrt{\kappa_1\kappa_2}E_{R1}E_{i2}\right)e^{-\frac{\alpha}{4}\frac{L}{4}-jk_n\frac{L}{4}} \tag{2.19}$$

The power output of the through port (P_{t1}) is written by

$$P_{t1} = (E_{t1})\cdot(E_{t1})^* = \left|x_1y_1E_{i1} + \left(jx_1x_2y_2\sqrt{\kappa_1}E_{R2}E_{R1}E_1 - x_1x_2\sqrt{\kappa_1\kappa_2}E_{R1}E_{i2}\right)e^{-\frac{\alpha}{4}\frac{L}{4}-jk_n\frac{L}{4}}\right|^2. \tag{2.20}$$

Similarly, from equations (2.5), (2.6), (2.16)-(2.18), the output optical field of the drop port (E_{t2}) is given by

$$E_{t2} = x_2y_2E_{i2} + jx_2\sqrt{\kappa_2}E_{R2}E_1e^{-\frac{\alpha}{4}\frac{L}{4}-jk_n\frac{L}{4}} \tag{2.21}$$

The power output of the drop port (P_{t2}) is expressed by

$$P_{t2} = (E_{t2})\cdot(E_{t2})^* = \left|x_2y_2E_{i2} + jx_2\sqrt{\kappa_2}E_{R2}E_1e^{-\frac{\alpha}{4}\frac{L}{4}-jk_n\frac{L}{4}}\right|^2. \tag{2.22}$$

In order to retrieve the required signals, we propose to use the add/drop optical multiplexing device with the appropriate parameters. This is given in the following details. The optical circuits of a *PANDA* ring resonator for the through port and drop port can be given by equations (2.20) and (2.22), respectively. The chaotic noise cancellation can be managed by using the specific parameters of the add/drop multiplexing device. The required signals can be retrieved by the specific users. κ_1 and κ_2 are the coupling coefficients of the add/drop filters, $k_n=2\pi/\lambda$ is the wave propagation number for in a vacuum, and the waveguide (ring resonator) loss is $\alpha = 5\times10^{-5}$ dBmm^{-1}. The fractional coupler intensity loss is $\gamma = 0.01$. In the case of the add/drop multiplexing device, the nonlinear refractive index is neglected. Figure 3 shows the schematic diagram of single-PANDA ring by using OptiFDTD commercial software. The dynamic pulse train is generated in z-direction of InGaAsP/InP waveguide with $n_0 = 3.34$ by using OptiFDTD as shown in Figure 2.4, (a) z = 0, (b) z = 0.88μm, (c) z = 1.62μm, (d) z = 2.74μm, (e) z = 3.30μm, (f) z = 3.44μm, (g) z = 5.72μm, (h) z = 8.10μm, and (i) z = 9.06μm, (j) z = 10.7 μm, (k) z = 10.0μm, and (l) z = 11.0μm. The results are obtained by the through (Th) and drop (Dr) ports as shown in Figure 2.5. We found that the output power at the drop port is higher than the through port, which means that the required and the transmitted signals are obtained, in which the delay signals within left and right nanoring are seen as shown in Figure 2.6.

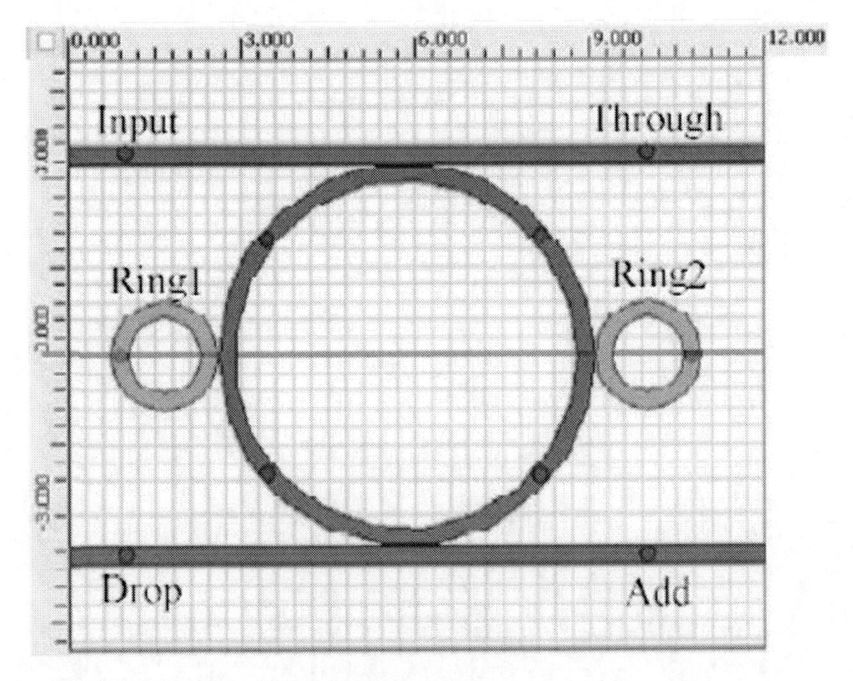

Figure 2.3. A schematic of single-PANDA ring with a 10 × 12μm^2 size, which is drawn by using the OptiFDTD commercial software.

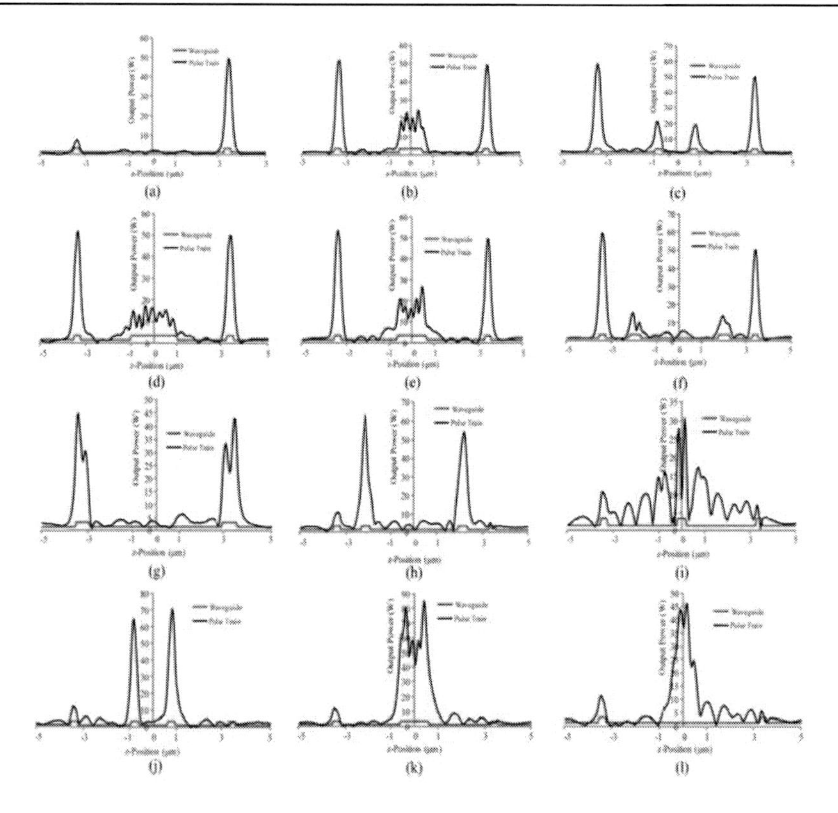

Figure 2.4. Dynamic pulse in z-direction of PANDA ring size $10 \times 12\ \mu m^2$ by using OptiFDTD, where (a) z = 0, (b) z = 0.88μm, (c) z = 1.62μm, (d) z = 2.74μm, (e) z = 3.30μm, (f) z = 3.44μm, (g) z = 5.72μm, (h) z = 8.10μm, and (i) z = 9.06μm, (j) z = 10.7 μm, (k) z = 10.0μm, and (l) z = 11.0μm.

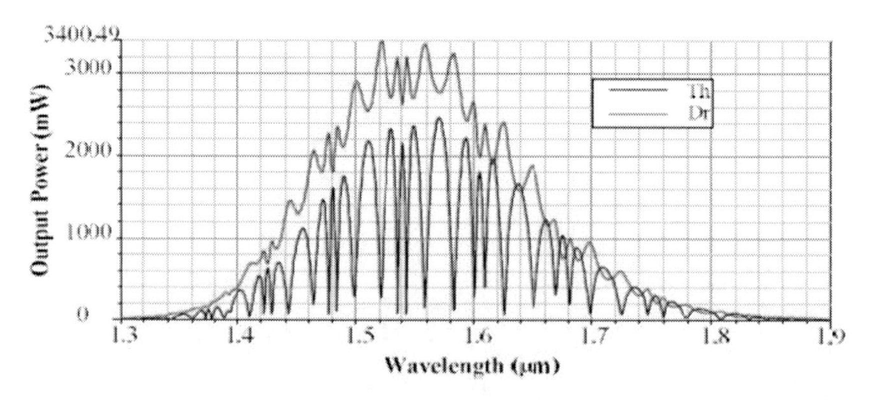

Figure 2.5. The simulation results obtained at the through port (Th) and drop port (Dr).

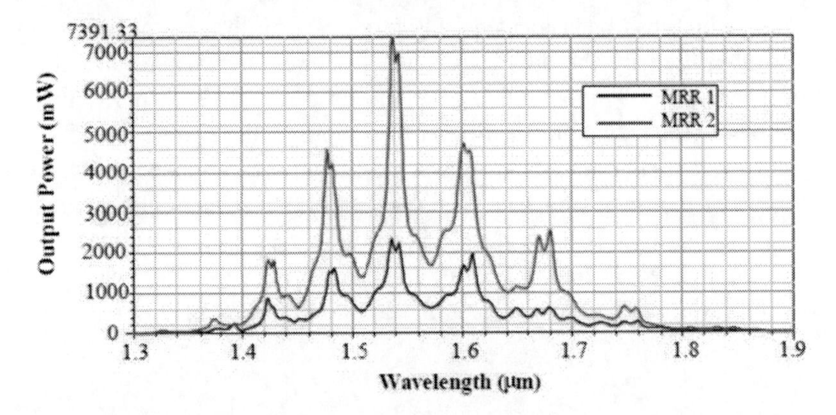

Figure 2.6. The simulation results generated at ring1 (MRR1) and ring2 (MRR2).

2.3. Dynamic Pulses Propagation

A schematic diagram of the PANDA ring resonator system [10-22] for dynamic pulse observation is designed and shown in Figure 2.7. In operation, to form the amplification part, a nanoring resonator is embedded within the add/drop optical filter in the system. The modulated Gaussian continuous wave (CW) with center wavelength (λ_0) at 1.55 μm, peak power at 50 mW is input into the system. The combination between the input and reflected light output can be seen as shown in Figure 2.8(a). The suitable ring parameters are used, for instance, ring radii $R_1=R_2=R_3=$ 0.80 μm, and the add/drop, R_{ad} = 3.2 μm. In order to make the system associate with the practical device [8, 9], the selected parameters of the system are fixed to n_0 = 3.34 (*InGaAsP/InP*), A_{eff} = 0.50 μm^2 and 0.1 μm^2 for a microring (add/drop filter) and the nanoring, respectively, α = 0.5 $dBmm^{-1}$, γ = 0.1. In this investigation, the coupling coefficient (kappa,κ) of the microring resonator is 0.8. The nonlinear refractive index of the microring used is $n_2=2.2 \times 10^{-17}$ m^2/W. In this case, the attenuation of light propagates within the system (i.e. wave guided) used is 0.5$dBmm^{-1}$. The wafer (device) length (z) is 20 μm, width = 10 μm.

2.4. Results and Discussion

The dynamic locations can be configured as followings; Ead: pulse propagation in the add/drop device and Er: pulse propagation in the nanoring resonator, which has been located in Figure 2.7. The through and drop ports are also identified by the add/drop filter structure. In practice, the required output signals are obtained and seen at the drop and through ports. In Figs. 2.8-2.11 show the results in the different locations in the double PANDA ring resonator system. The maximum power of 120 mW is obtained as shown in Figure 2.4(b) that is in the nanoring (Er1), while the maximum number of peaks of 18 is seen in Figure 2.11(b). The three dimension (3-D) image of the dynamic modulated Gaussian CW is as shown in Figure 2.12, whereas the maximum power within the nanoring of 120 mW is obtained.

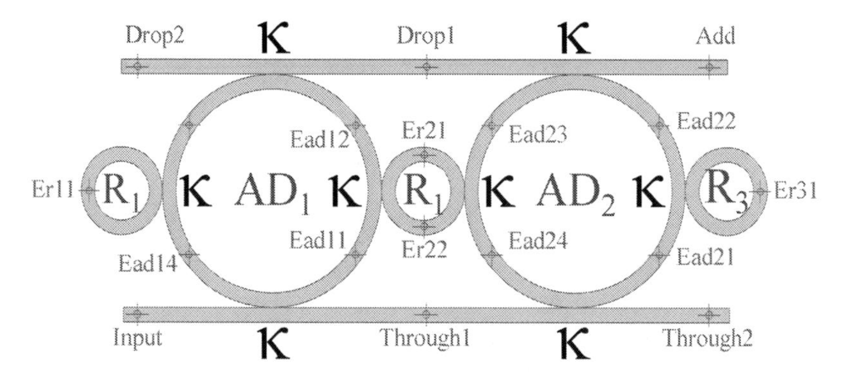

Figure 2.7. A schematic of a double PANDA ring resonators with the dynamic locations, where ADs: add/drop filter.

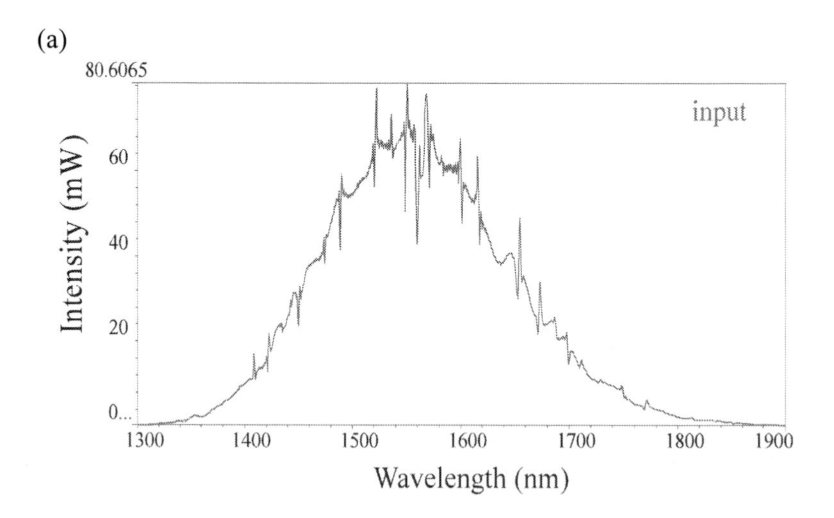

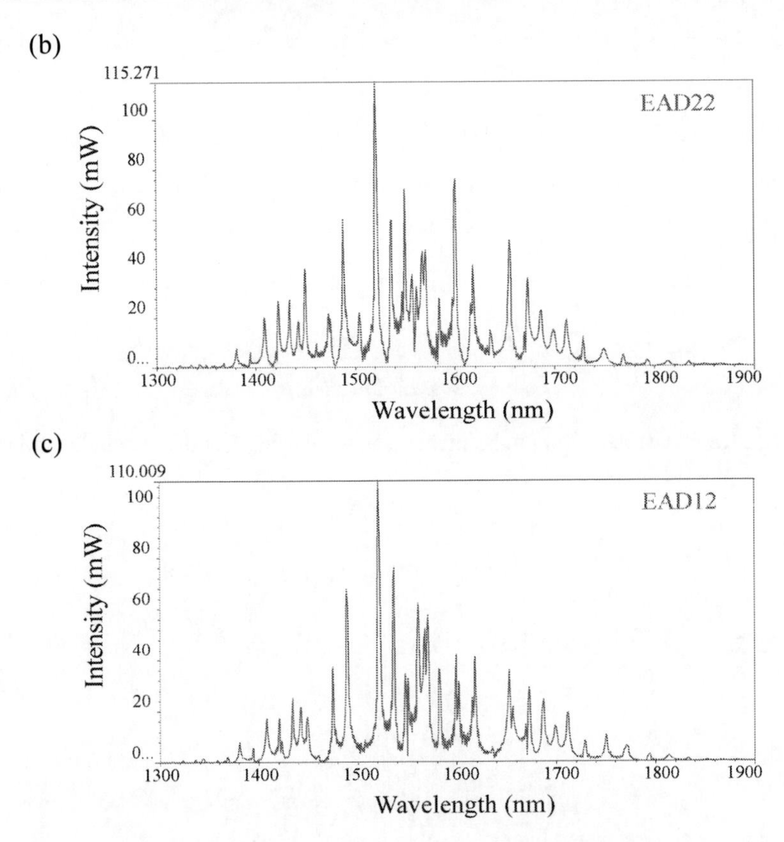

Figure 2.8. Results of the output light intensity and wavelength at the certain location at (a) an input pulse, (b) EAD22: Ead 22 and (c) EAD12: Ead12.

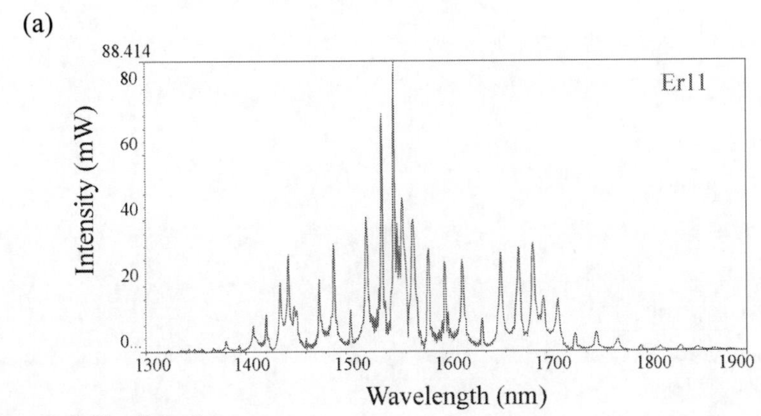

Figure 2.9 (Continues)

(b)

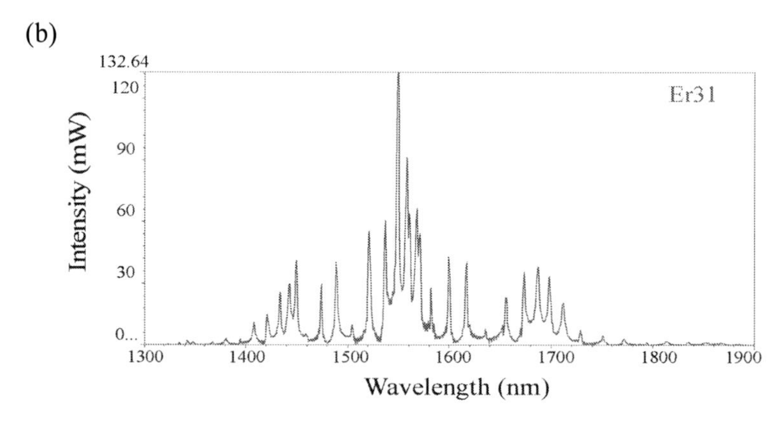

Figure 2.9. Results of the output light intensity and wavelength at the certain location at (a) Er 11 and (b) Er 31.

(a)

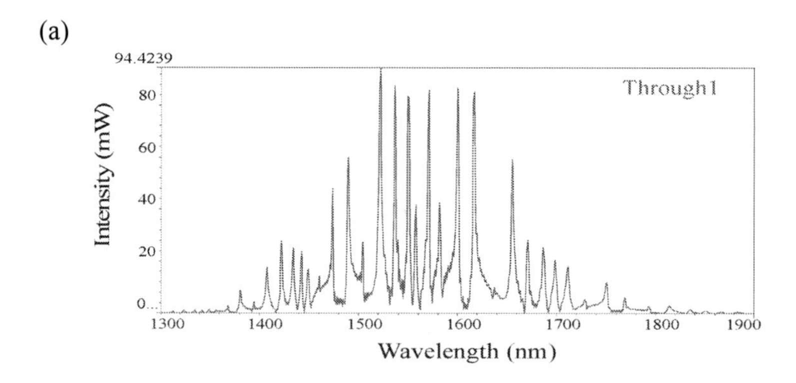

(b)

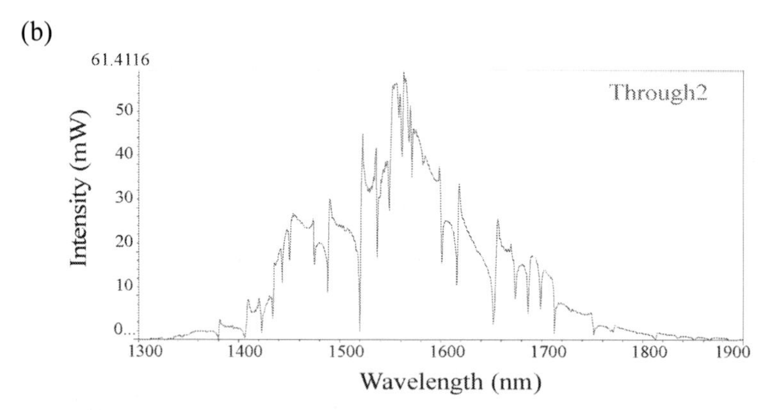

Figure 2.10. Results of the output light intensity and wavelength at the certain location at (a) Through 1 and (b) Through 2.

(a)

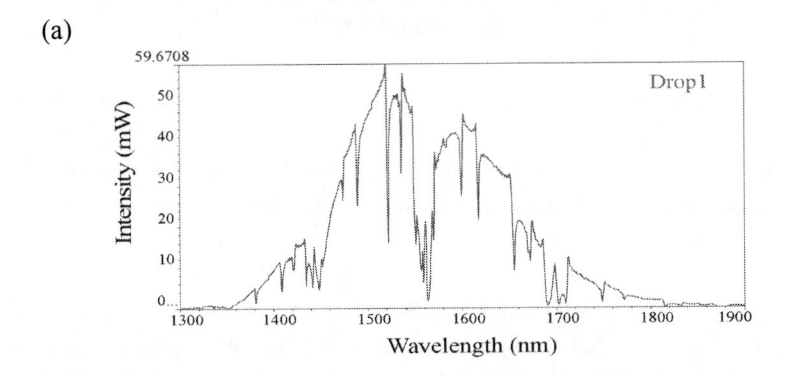

(b)

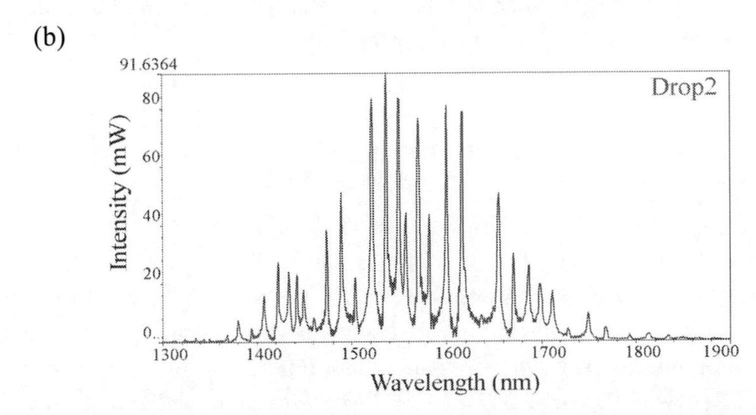

Figure 2.11. Results of the output light intensity and wavelength at the certain location at (a) Drop 1 and (b) Drop 2.

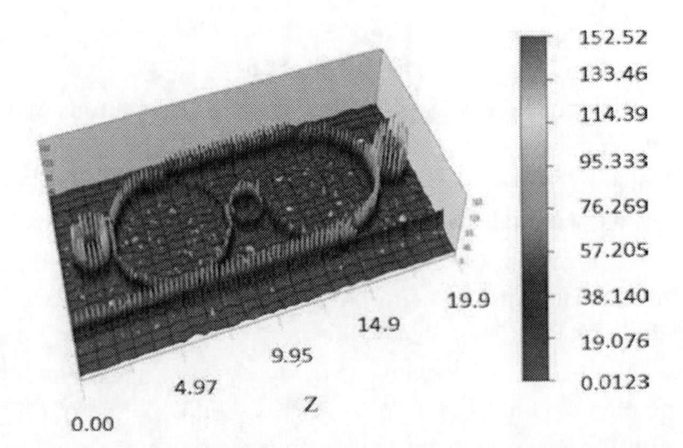

Figure 2.12. The 3-D dynamic graphic results obtained using the OPTI-WAVE PROGRAMMING.

2.5. Conclusion

We have shown the dynamic behaviors of pulses propagation within the double PANDA ring resonators that can be performed by using two nanoring resonators incorporating into the add/drop multiplexer. By using the OPTI-WAVE PROGRAMMING and the reasonable input parameters, the dynamic behaviors can be controlled and the required output pulses obtained. In this case, the dynamic behavior can be controlled and used for the desired applications. The maximum number of peaks (counts) of 18 is obtained, which is available for multi light source or wavelength enhancement application, moreover, the output amplification is also obtained, where the maximum power within the nanoring of 120 mW is obtained.

References

[1] D. Deng and Q. Guo, "Ince-Gaussian solitons in strongly nonlocal nonlinear media", *Opt. Lett.*, 32(2007)3206-3208.

[2] P.P. Yupapin and W. Suwancharoen, "Chaotic signal generation and cancellation using a microring resonator incorporating an optical add/drop multiplexer, *Opt. Commun.*, 280(2007)343-350.

[3] A. Hasegawa, "*Massive WDM and TDM Soliton Transmission Systems,*" Kluwer Academic Publishers, Boston, 2000.

[4] G. P. Agarwal, "*Nonlinear Fiber Optics*", 4th edition, Academic Press, New York, 2007.

[5] Yu. A. Simonov and J. A. Tjon, "Soliton-soliton interaction in confining models," *Phys. Lett.*, B 85(1979)380-384.

[6] Takeshi Iizuka and Yuri S. Kivshar, "Optical gap solitons in nonresonant quadratic media, *Phys. Rev. E* 59(1999)7148 - 7151.

[7] P.P. Yupapin, P. Saeung and C. Li, "Characteristics of complementary ring-resonator add/drop filters modeling by using graphical approach," *Opt. Commun.*, 272(2007)81-86.

[8] Q. Xu and M. Lipson, "*All-optical logic based on silicon micro-ring resonators,*" 15(3)(2007)924-929.

[9] Y. Kokubun, Y. Hatakeyama, M. Ogata, S. Suzuki, and N. Zaizen, "Fabrication technologies for vertically coupled micro ring resonator with multilevel crossing busline and ultracompact-ring radius," *IEEE J. Sel. Top. Quantum Electron.* 11(2005)4-10.

[10] K. Uomwech, K. Sarapat, and P. P. Yupapin, "Dynamic modulated Gaussian pulse propagation within the double PANDA ring resonator system," Microw. *Opt. Techn. Lett., vol.* 52, no. 8, pp. 1818-1821, Aug 2010.

[11] K. Sarapat, N. Sangwara, K. Srinuanjan, P.P. Yupapin and N. Pornsuwancharoen, "Novel dark-bright optical solitons conversion system and power amplification," *Opt. Eng.*, vol.48, pp. 045004, 2009.

[12] OptiFDTD by OptiWave Corp. ©, ver. 8.0, *single license (KMITL),* 2008.

[13] T. Phatharaworamet, C. Teeka, R. Jomtarak, S. Mitatha, and P. P. Yupapin, "Random Binary Code Generation Using Dark-Bright Soliton Conversion Control Within a PANDA Ring Resonator," *J. Lightw. Techn.,* vol. 28, no. 19, pp. 2804-2809, 2010.

[14] N. Suwanpayak, M. A. Jalil, C. Teeka, J. Ali, and P. P. Yupapin, " Optical vortices generated by a PANDA ring resonator for drug trapping and delivery applications," *Biomed. Opt. Express*, vol. 2, no. 1, pp. 159-168, January 2011.

[15] P. P. Yupapin, N. Suwanpayak, B. Jukgoljun, and C. Teeka, "Hybrid Transceiver using a PANDA Ring Resonator for Nano Communication," *Physics Express*, vol. 1(1), pp. 1-8, Feb. 2011.

[16] M. Tasakorn, C. Teeka, R. Jomtarak, and P. P. Yupapin, "Multitweezers generation control within a nanoring resonator system," *Optical Engineering,* vol. 49, no. 7, 075002 (July 2010).

[17] B. Jukgoljun, N. Suwanpayak, C. Teeka, and P.P. Yupapin, "Hybrid Transceiver and Repeater using a PANDA Ring Resonator for Nano Communication," *Optical Engineering*, vol. 49, no. 12, 125003 (2010).

[18] P. Youplao, T. Phattaraworamet, S. Mitatha, C. Teeka, and P. P. Yupapin, "Novel optical trapping tool generation and storage controlled by light," *Journal of Nonlinear Optical Physics and Materials*, vol. 19, no. 2, pp. 371–378, (June 2010).

[19] K. Sarapat, N. Sangwara, K. Srinuanjan, P.P. Yupapin and N. Pornsuwancharoen, "Novel dark-bright optical solitons conversion system and power amplification," *Opt. Eng.*, 48, 045004-1-5(2009).

[20] S. Mitatha, N. Chaiyasoonthorn, P. P. Yupapin: "Dark-bright optical solitons conversion via an optical add/drop filter," *Microw. Opt. Technol. Lett.,* 51, 2104-2107(2009).

[21] T. Threepak, X. Luangvilay, S. Mitatha and P.P. Yupapin, "Novel quantum-molecular transporter and networking via a wavelength router," *Microw. Opt. Technol. Lett.*, 52(6), 1353-1357 (2010).

[22] B. Piyatamrong, K. Kulsirirat, W. Techithdeera, S. Mitatha, P.P. Yupapin, "Dynamic potential well generation and control using double resonators incorporating in an add/drop filter. *Mod Phys Lett B.* 2010;24:3071–3082.

Chapter 3

NEW COMMUNICATION BAND

3.1. INTRODUCTION

Optical soliton is recognized as the powerful light source for long distance link in optical communication. However, the pumping system is required before the soliton being generated. For simplicity, a Gaussian soliton is recommended to form the soliton instead of the pumping soliton. Moreover, one interesting aspect of the Gaussian soliton is that the non-dispersive soliton can be realized by using the 1,300 nm light source, which can be obtained by using the Gaussian soliton. Therefore, the use of a Gaussian soliton becomes a very attractive tool in the area of soliton investigation, whereas the simple system arrangement can be used to form the soliton behavior within the medium for various investigations. Many research works have reported in use of a Gaussian pulse in both theoretical and experimental works [1-4]. Recently, the interesting aspect of light pulse propagating within a nonlinear microring device has been reported [5], where the transfer function of the output at resonant condition is derived and studied. They found that the broad spectrum of light pulse can be transformed to the discrete pulses. An optical soliton has been employed as a powerful laser pulse, which can be used to enlarge the optical bandwidth when propagating within the nonlinear microring resonator [6, 7]. Moreover, the superposition of self-phase modulation (SPM) soliton pulses, where either bright or dark [8] solitons can generate the large output power. For further reading, many earlier works of soliton applications in either theory or experimental works are found in a soliton application book by Hasegawa et al [9]. Many of the soliton related concepts in fiber optic are discussed by Agrawal [10]. The problems of soliton-soliton interactions [11], collision [12], rectification [13] and dispersion management [14] are required to solve and address. Therefore, in this work we are looking for a common laser source that can be used to extend the used/installed wavelength band in the public network, especially, with the broad center wavelengths within the range from 400-1,500 nm are used. Therefore, the use of a Gaussian pulse to form a broad wavelength bands within a microring resonator system is recommended, where the other problems such as are output power and signal collision can be solved. One of the results has shown that by using the center wavelength at 1,300 nm with suitable parameters, the Gaussian soliton generation is plausible. By using the suitable microring parameters, most of the results have shown that the optical signals, i.e. Gaussian pulse can be amplified within the nonlinear ring resonator system.

The demand of communication channels and network capacity has been increased significantly for three decades, however, up to now, the large user demand remains. Therefore, the searching of new techniques is needed, which is focused on the communication channel and network capacity. Recently, Pornsuwancharoen et al [15] have reported the very interesting result of the technique that can be used to fulfill the large demand. They have shown that the signal bandwidth can be stretched and compressed by using the nonlinear micro ring system [16-18]. By using such a scheme, the increasing in communication channels using soliton communication is plausible. Furthermore, the long distance communication link is also available. However, several problems are required to solve and address, for instance, the problem of soliton-soliton interaction and collision [19], and the waveguide structure that the broadband soliton can be confined [20]. In this letter, we propose the technique that can be used to generate the new soliton communication bands (wavelength bands), whereas the common soliton pulse, i.e. a soliton source is at the center wavelength of 1.55 μm. The soliton bands at the required center wavelengths can be stored [21] and filtered by using the add/drop filter [19]. In application, the use of super dense wavelength multiplexing, with the long distance link is available. Furthermore, the personnel channel and network may be plausible due to the very available bandwidths. However, the problem of the soliton interaction and collision is required to solve, which can be avoided by the specific free spectrum range design [19].

Wireless communication technology has become a part of human life, which is recognized as the convenient tools in the world society. Up to date, the merging communication system has become more realistic and available. The wireless network, whereas the demand has been increased rapidly.

Generally, the wireless network communications performed by using radio frequency electromagnetic wave to share information and resources between wireless devices; such as mobile terminal, pocket size PCs, hand-held PCs, laptops, cellular phone, PDAs, wireless sensors, and satellite receivers. Digital signal processing (DSP) is ideas of software define ratio (SDR) [22] mechanism, broadcast message between transmitter and receiver by broadcast channel. Broadcast channel is the basic form of communication in all wireless system by medium access control (MAC) and CSMA/CA protocol. The wireless operates by two type modes are infrastructure-based and Mobile Ad Hoc Network (MANET). The MANET formed dramatically through the cooperation and self organizations of mobile nodes; connect via wireless link, no centralized administrator and free to move randomly. MANET used IEEE 802.11 standard and CSMA/CA in this standard used to provide collision avoidance and congestion control. Two mechanisms for performs in MANET are broadcast protocol [23-25]; one available ad hoc node attempts to broadcast message to all participation nodes by broadcast mechanisms and routing protocols[26]; search or find between the pair nodes by some mechanisms such as DSDV, CGSR, WRP, GSR, OLSR, FSR, LAN-MAR, HRS, DSR, AODV, TORA, ABR, and SSR. Normally, MANET link by radio frequency and used a channel for communicate with other participant nodes in a coverage area by used CSMA/CA protocol to solve hidden and expose problems, in other way, these problem can resolve by some method such as multichannel communication [27]. In case out of coverage area, MANET communicates with other coverage via the relay node, this link is the platform of multi-hop network. The performance of communication for Ad Hoc network contain with many factors such as the bandwidth of channel, number of node, the velocity of node, and the technique for communication management. Group [28] and cluster-based [29] accompany the mobile device such as processors, memory, and I/O devices. Ad Hoc overlay network [30] is the virtual network for resources management in Ad hoc such as dissemination, discovery, or other process. The hybrid network [31], combine various type of technology to wireless capability, wire network, wireless network, GPS, and CDMA [32]. The diversity mechanism [33], transmit more than one channels by using antenna array and received best channel for data transmissions. This research we propose the new dedicate intermediary link between nodes in MANET system by using the dense wavelength division multiplexing (DWDM) by point-to-point fashion. Every MANET communicates together with participant nodes by direct one-to-one link or by via relay node with THz antenna [34].

3.2. New Communication Bands Using Gaussian Pulse

Light from a monochromatic light source is launched into a ring resonator with constant light field amplitude (E_0) and random phase modulation as shown in Figure 3.1, which is the combination of terms in attenuation (α) and phase(ϕ_0) constants, which results in temporal coherence degradation. Hence, the time dependent input light field (E_{in}), without pumping term, can be expressed as

$$E_{in}(t) = E_0 e^{-\alpha L + j\phi_0(t)}. \tag{3.1}$$

where L is a propagation distance(waveguide length).

We assume that the nonlinearity of the optical ring resonator is of the Kerr-type, i.e., the refractive index is given by

$$n = n_0 + n_2 I = n_0 + (\frac{n_2}{A_{eff}})P, \tag{3.2}$$

where n_0 and n_2 are the linear and nonlinear refractive indexes, respectively. I and P are the optical intensity and optical power, respectively. The effective mode core area of the device is given by A_{eff} . For the microring and nanoring resonators, the effective mode core areas range from 0.10 to 0.50 μm^2 [12]

When a Gaussian pulse is input and propagated within a fiber ring resonator, the resonant output is formed, thus, the normalized output of the light field is the ratio between the output and input fields ($E_{out}(t)$ and $E_{in}(t)$) in each roundtrip, which can be expressed as [13]

$$\left|\frac{E_{out}(t)}{E_{in}(t)}\right|^2 = (1-\gamma)\left[1 - \frac{(1-(1-\gamma)x^2)\kappa}{(1-x\sqrt{1-\gamma}\sqrt{1-\kappa})^2 + 4x\sqrt{1-\gamma}\sqrt{1-\kappa}\sin^2(\frac{\phi}{2})}\right] \tag{3.3}$$

Equation (3.3) indicates that a ring resonator in the particular case is very similar to a Fabry-Perot cavity, which has an input and output mirror with a field reflectivity, $(1-\kappa)$, and a fully reflecting mirror. κ is the coupling coefficient, and $x = \exp(-\alpha L/2)$ represents a roundtrip loss coefficient, $\phi_0 = kLn_0$ and $\phi_{NL} = kL(\frac{n_2}{A_{eff}})P$ are the linear and nonlinear phase shifts, $k = 2\pi/\lambda$ is the wave propagation number in a vacuum. Where L and α are a waveguide length and linear absorption coefficient, respectively. In this work, the iterative method is introduced to obtain the results as shown in equation (3.3), similarly, when the output field is connected and input into the other ring resonators.

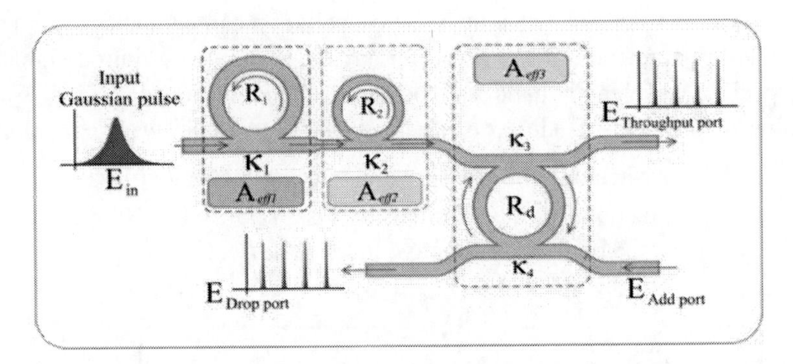

Figure 3.1. A schematic of a Gaussian soliton generation system, where R_s: ring radii, κ_s: coupling coefficients, R_d: an add/drop ring radius, A_{eff}s: Effective areas.

The input optical field as shown in equation (3.1), i.e. a Gaussian pulse, is input into a nonlinear microring resonator. By using the appropriate parameters, the chaotic signal is obtained by using equation (3.3). To retrieve the signals from the chaotic noise, we propose to use the add/drop device with the appropriate parameters. This is given in details as followings. The optical outputs of a ring resonator add/drop filter can be given by the equations (3.4) and (3.5).

$$\left|\frac{E_t}{E_{in}}\right|^2 = \frac{(1-\kappa_1) - 2\sqrt{1-\kappa_1}\cdot\sqrt{1-\kappa_2}e^{-\frac{\alpha}{2}L}\cos(k_n L) + (1-\kappa_2)e^{-\alpha L}}{1+(1-\kappa_1)(1-\kappa_2)e^{-\alpha L} - 2\sqrt{1-\kappa_1}\cdot\sqrt{1-\kappa_2}e^{-\frac{\alpha}{2}L}\cos(k_n L)} \tag{3.4}$$

and

$$\left|\frac{E_d}{E_{in}}\right|^2 = \frac{\kappa_1\kappa_2 e^{-\frac{\alpha}{2}L}}{1+(1-\kappa_1)(1-\kappa_2)e^{-\alpha L} - 2\sqrt{1-\kappa_1}\cdot\sqrt{1-\kappa_2}e^{-\frac{\alpha}{2}L}\cos(k_n L)} \tag{3.5}$$

where E_t and E_d represents the optical fields of the throughput and drop ports respectively. The transmitted output can be controlled and obtained by choosing the suitable coupling ratio of the ring resonator, which is well derived and described by reference [16]. Where $\beta = kn_{eff}$ represents the propagation constant, n_{eff} is the effective refractive index of the waveguide, and the circumference of the ring is $L = 2\pi R$, here R is the radius of the ring. In the following, new parameters will be used for simplification, where $\phi = \beta L$ is the phase constant. The chaotic noise cancellation can be managed by using the specific parameters of the add/drop device, which the required signals at the specific wavelength band can be filtered and retrieved. $K_1$and K_2 are coupling coefficient of add/drop filters, $k_n = 2\pi / \lambda$ is the wave propagation number for in a vacuum, and the waveguide (ring resonator) loss is α = 0.5 $dBmm^{-1}$. The fractional coupler intensity loss is γ = 0.1. In the case of add/drop device, the nonlinear refractive index is neglected.

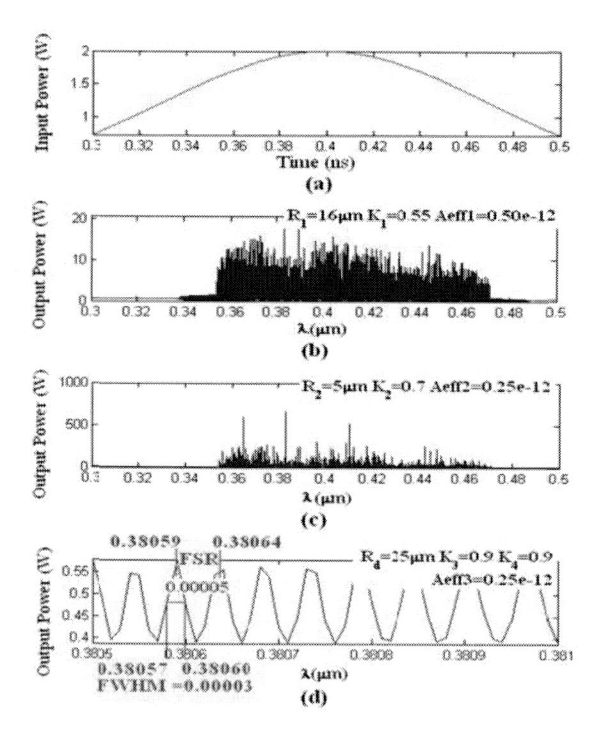

Figure 3.2. Result of the spatial pulses with center wavelength at 0.40 μm, where (a) the Gaussian pulse, (b) large bandwidth signals, (c) large amplified signals, (d) filtering and amplifying signals from the drop port.

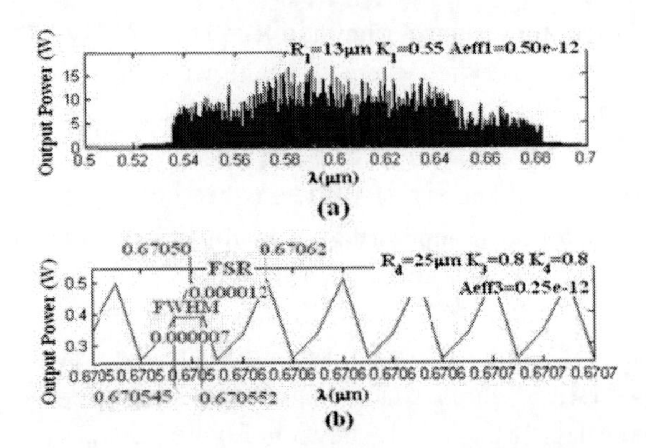

Figure 3.3. Result of the spatial pulses with center wavelength at 0.60 μm, where (a) large bandwidth signals, (b) filtering and amplifying signals from the drop port.

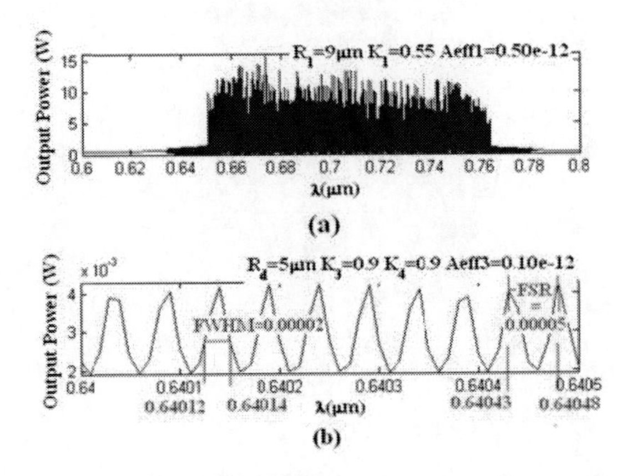

Figure 3.4. Result of the spatial pulses with center wavelength at 0.70 μm, where (a) large bandwidth signals, (b) filtering and amplifying signals from the drop port.

From Figure 3.1, in principle, light pulse is sliced to be the discrete signal and amplified within the first ring, where more signal amplification can be obtained by using the smaller ring device (second ring). Finally, the required signals can be obtained via a drop port of the add/drop filter. In operation, an optical field in the form of Gaussian pulse from a laser source at the specified center wavelength is input into the system. From Figure 3.2, the Gaussian pulse with center wavelength (λ_0) at 0.40 μm, pulse width (Full Width at Half Maximum, FWHM) of 20 ns, peak power at 2 W is input into the system as shown in Figure 3.2(a). The large bandwidth signals can be seen within the first microring device, and shown in Figure 3.2(b). The suitable ring parameters are used, for instance, ring radii R_1= 16.0 μm, R_2= 5.0 μm, and R_d= 25.0 μm. In order to make the system associate with the practical device [35], the selected parameters of the system are fixed to n_0 = 3.34 (InGaAsP/InP), A_{eff} = 0.50 μm^2 and 0.25 μm^2 for a microring and add/drop ring resonator, respectively, α = 0.5 $dBmm^{-1}$, γ = 0.1. In this investigation, the coupling coefficient (kappa, κ) of the microring resonator is ranged from 0.55 to 0.90. The nonlinear refractive index of the microring used is n_2=2.2 x 10^{-17} m^2/W. In this case, the attenuation of light propagates within the system (i.e. wave guided) used is 0.5$dBmm^{-1}$. After light is input into the system, the Gaussian pulse is chopped (sliced) into a smaller signal spreading over the spectrum due to the nonlinear effects[5], which is shown in Figure 3.2(a).

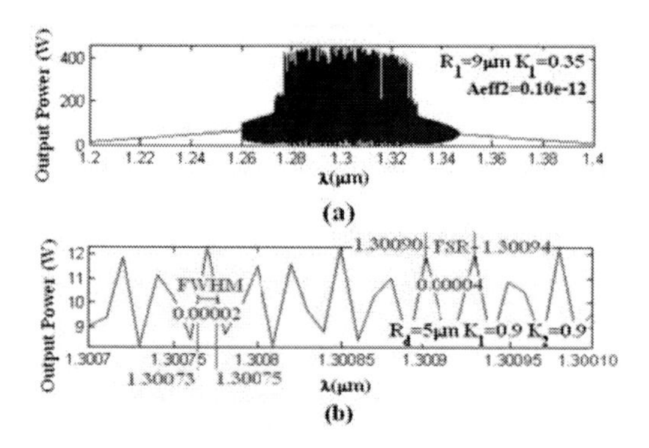

Figure 3.5. Result of the spatial pulses with center wavelength at 1.30 μm, where (a) large bandwidth signals, (b) filtering and amplifying signals from the drop port.

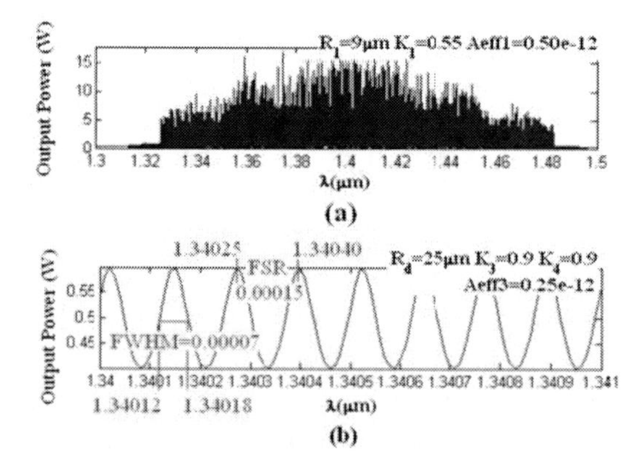

Figure 3.6. Result of the spatial pulses with center wavelength at 1.40 μm, where (a) large bandwidth signals, (b) filtering and amplifying signals from the drop port.

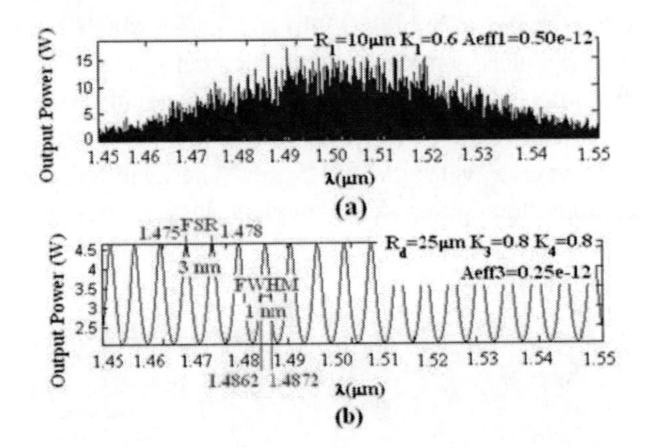

Figure 3.7. Results of the spatial pulses with center wavelength at 1.50 µm, where (a) large bandwidth signals, (b) filtering and amplifying signals from the drop port.

The large bandwidth signal is generated within the first ring device. In applications, the specific input or out wavelengths can be used and generated. For instance, the different center wavelengths of the input pulse can be ranged from 0.40-1.5 µm as shown in Figs. 3.2-3.7, where the suitable parameters are used and shown in the figures.

3.3. New Communication Bands Using Soliton

In operation, the large bandwidth signal within the micro ring device can be generated by using a common soliton pulse input into the nonlinear micro ring resonator. This means that the broad spectrum of light can be generated after the soliton pulse is input into the ring resonator system. The schematic diagram of the proposed system is as shown in Figure 3.8. A soliton pulse with 50 ns pulse width, peak power at 2 W is input into the system. The suitable ring parameters are used, for instance, ring radii R_1= 15.0 µm, R_2= 10.0 µm, R_3= R_s=5.0 µm and R_5=R_d= 20.0 µm. In order to make the system associate with the practical device [36], the selected parameters of the system are fixed to λ_0 = 1.55 µm, n_0 = 3.34 (InGaAsP/InP), A_{eff} = 0.50, 0.25 µm^2 and 0.10 µm^2 for a micro ring and nano ring resonator, respectively, α = 0.5 dBmm^{-1}, γ = 0.1. The coupling coefficient (kappa, κ) of the micro ring resonator ranged from 0.1 to 0.95. The nonlinear refractive index is n_2=2.2×10^{-13} m^2/W. In this case, the wave guided loss used is 0.5dBmm^{-1}. The input soliton pulse is chopped (sliced) into the smaller signals spreading over the spectrum (i.e.

broad wavelength) as shown in Figs. 3.9(b) and 3.9(b), which is shown that the large bandwidth signal is generated within the first ring device. The biggest output amplification is obtained within the nano-waveguides (rings R_3 and R_4) as shown in Figs. 3.9(d) and 3.9(e), whereas the maximum power of 10 W is obtained at the center wavelength of 1.5 μm. The coupling coefficients are given as shown in the figures. The coupling loss is included due to the different core effective areas between micro and nano ring devices, which is given by 0.1dB.

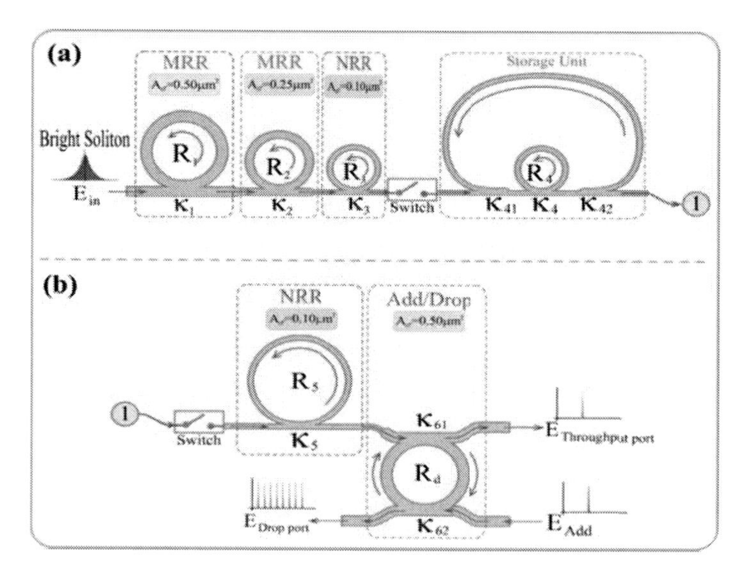

Figure 3.8. A broadband generation system, (a) a broadband source generation and a storage unit, (b) a soliton band selector, where R_s: ring radii, κ_s: coupling coefficients, κ_{41}, κ_{42}: coupling losses, κ_{61} and κ_{61} are the add/drop coupling coefficients.

We have shown that a large bandwidth of the optical signals with the specific wavelength can be generated within the micro ring resonator system as shown in Figure 3.8. The amplified signals with broad spectrum can be generated, stored and regenerated within the nano-waveguide. The maximum stored power of 10 W is obtained as shown in Figure 3.9(d) and 3.9(e), where the average regenerated optical output power of 4 W is achieved via and a drop port of an add/drop filter as shown in Figure 3.9(h)-3.9(k), which is a broad spectra of light cover the large bandwidth as shown in Figure 3.9(g). However, to make the system being realistic, the waveguide and connection losses are required to address in the practical device, which may be affected the signal amplification.

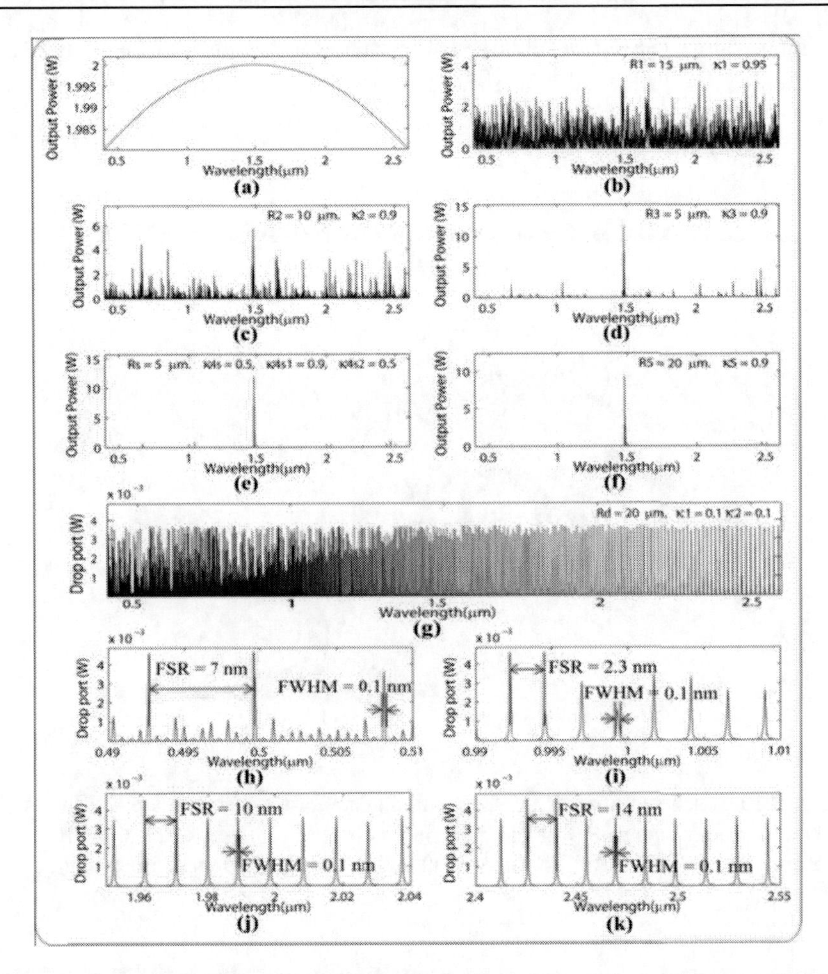

Figure 3.9. A soliton band with center wavelength at 1.5 μm, where (a) input soliton, (b) ring R_1, (c) ring R_2, (d) ring R_3, (e) storage ring(R_s), (f) ring R_5, (g) drop port output signals. The output of different soliton bands (center wavelength) are as shown, where (h) 0.51 μm, (i) 0.98 μm, (j) 1.99 μm and (k) 2.48 μm.

The storage light pulse within a storage ring (R_s or R_4) is achieved, which has also been reported by Ref. [21]. In applications, the increasing in communication channel and network capacity can be formed by using the different soliton bands (center wavelength) as shown in Figure 3.9, where 3.9(h) 0.51 μm, 3.9(i) 0.98 μm , 3.9(j) 1.48 μm and 3.9(k) 2.46 μm are the generated center wavelengths of the soliton bands. The selected wavelength center can be performed by using the designed add/drop filter, where the required spectral width (Full Width at Half Maximum, FWHM) and free spectrum range (FSR) are obtained, the channel spacing and bandwidth are represented by FSR and FWHM, respectively, for instance, the FSR and FWHM of 2.3 nm and 100 pm are obtained as shown in Figure 3.9(i).

3.4. Ad Hoc Network Using Frequency Enhancement

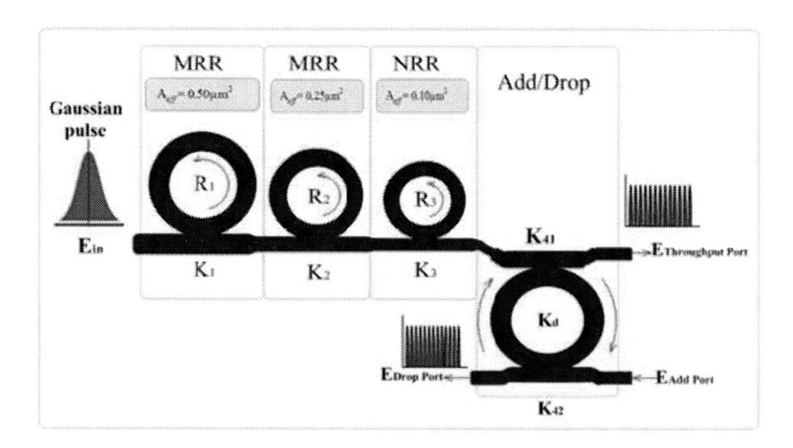

Figure 3.10. A schematic of a Gaussian soliton generation system, where R_s: ring radii, κ_s: coupling coefficients, R_d: an add/drop ring radius, A_{eff}: Effective areas, MRR: Microring resonator, NRR: Nanoring resonator, K_{42} and K_{42} are add/drop coupling coefficients.

High Capacity Ad Hoc Network Using Wireless Link

MANET is an autonomous node and independent resources management, majority used a channel for link all nodes by using CSMA/CA to access management. In this paper, we propose new platform for link wireless node by using a link per node (1-1), show in figure 3.10.

From Figs. 3.11, depict the Ad Hoc link model, (a) nodes A, B, and C can communicate with all other nodes or in coverage, in this cast A link to B directly, B link to C directly, and A link to C by directly. From fig 3.11(b), all nodes not in coverage, node A, B, and C are in coverage, nodes C and D are in coverage, and nodes D, C, and E are in coverage. From fig 3.11(c), show diagram for link by 1-1 of node A, node A can link to node B directly, node A can link to node C directly, node A can link to node B directly, but node A cannot link to nodes E and F directly due to out of coverage. In this case, we propose virtual direct link by using relay node, node A link to node E and F by used relay nodes C and D. Node A link to node E by using a channel via relay node C relay node D and in this case node A use four cannel for link in a time.

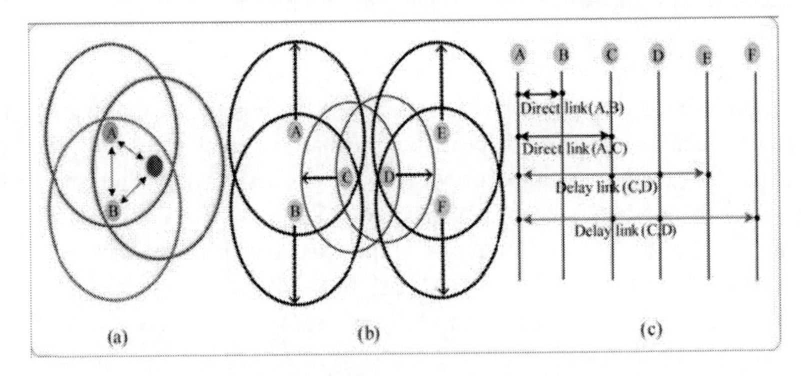

Figure 3.11. (a) Link 1-1 by direct node, (b) Link 1-1 via relay node, and (c) Diagram for node A link to all node (B, C, D, E, and F).

From Figs. 3.10 and 3.11, in principle, light pulse is sliced to be the discrete signal and amplified within the first ring, where more signal amplification can be obtained by using the smaller ring device (second ring). Finally, the required signals can be obtained via a drop port of the add/drop filter. In operation, an optical field in the form of Gaussian pulse from a laser source at the specified center wavelength (frequency) is input into the system. In practice, the maximum frequency that can be confined within the optical waveguide has been increased by using the composite of materials known as meta-materials [20], which is shown that the wavelength close to few mm (THz region) can be confined within the waveguide. In operation, light pulse is sliced to be the discrete signal and amplified within the first ring, where more signal amplification can be obtained by using the smaller ring device (second ring) as shown in Figure 3.10. Finally, the required signals can be obtained via a drop port of the add/drop filter. An optical field in the form of Gaussian pulse from a laser source at the specified center frequency is input into the system. From Figure 3.12, the Gaussian pulse with center frequency (f_0) at 3.0 THz, pulse width (Full Width at Half Maximum, FWHM) of 20 ns, peak power at 2 W is input into the system as shown in Figure 3.13(a). The large bandwidth signals can be seen within the first microring device, and shown in Figure 3.13(b). The suitable ring parameters are used, for instance, ring radii

R_1= 15.0 μm, R_2=R_3= 9.0 μm, and R_d= 50.0 μm. In order to make the system associate with the practical device [36, 37], the selected parameters of the system are fixed to n_0 = 3.34 (InGaAsP/InP), A_{eff} = 0.50 μm^2 and 0.25 μm^2 for a microring and add/drop ring resonator, respectively, α = 0.5 dBmm^{-1}, γ = 0.1. In this investigation, the coupling coefficient (kappa, κ) of the microring resonator is ranged from 0.10 to 0.96. The nonlinear refractive index of the microring used is n_2=2.2 x 10^{-17} m^2/W.

In this case, the attenuation of light propagates within the system (i.e. wave guided) used is 0.5dBmm^{-1}. After light is input into the system, the Gaussian pulse is chopped (sliced) into a smaller signal spreading over the spectrum due to the nonlinear effects[5], which is shown in Figure 3.13(b). The large bandwidth signal is generated within the first ring device. In applications, the specific input or output frequencies can be used and generated, where the suitable parameters are used and shown in the figures. The similar manner is as shown in Figs. 3.14-3.16, where the different parameters are the R_d radii and coupling coefficients, where the small FSR is obtained. In Figure 3.14, results of the THz frequency band with the center frequency at 3 THz, where (a) the input Gaussian pulse, (b) the large bandwidth signal, (c) the filtering and amplifying signals, (d) output frequency band, (e) and (f)are the drop port signals, (g) and (h)are the through port signals.

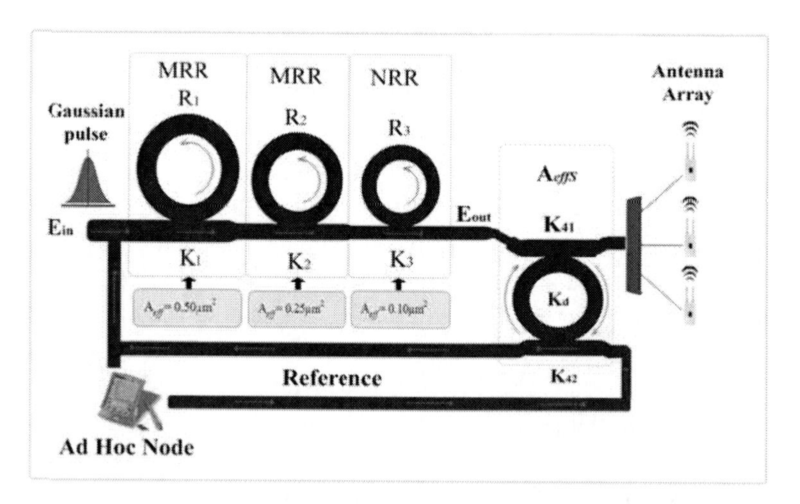

Figure 3.12. Ad Hoc wireless link model, where R_s: ring radii, κ_s: coupling coefficients, R_d: an add/drop ring radius, A_{eff}: Effective areas, MRR: Microring resonator, NRR: Nanoring resonator, K_{42} and K_{42} are add/drop coupling coefficients.

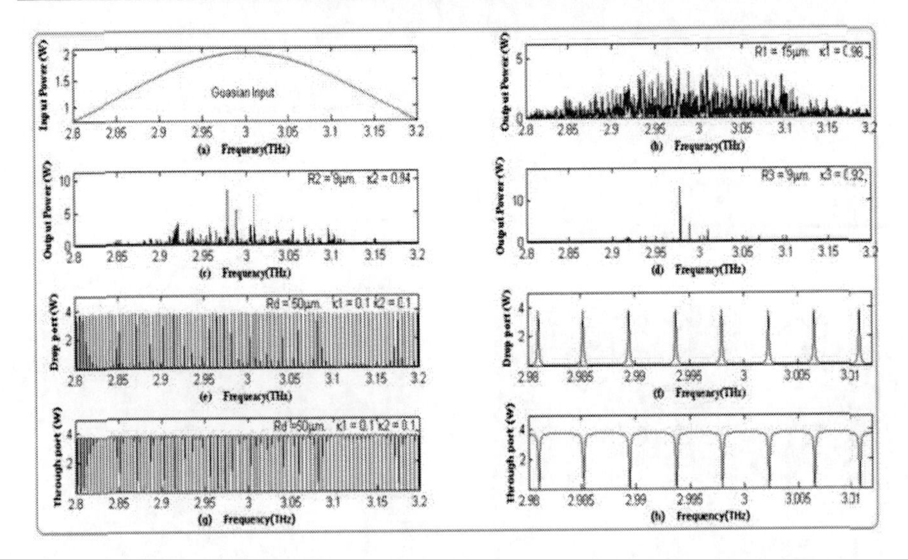

Figure 3.13. Results of the THz frequency band with the center frequency at 3THz, where (a) the input Gaussian pulse, (b) the large bandwidth signal, (c) the filtering and amplifying signals, (d) output frequency band, (e) and (f)are the drop port signals, (g) and (h)are the through port signals.

In application, the use of the generated frequency band for up-down link for wireless link is employed. Furthermore, there are several frequency bands available as shown in Figure 3.14-3.16, which can be brought the high capacity channels and multi switching system in the MANET system for direct link between mobile nodes or link via the relay nodes. By using the propose design as shown in Figure 3.12, the extended light source frequency bands can be used for multi frequency switching, which can be used employed via the simultaneous up link and down link system [38], the higher channel capacity can also be obtained by using FSR modification and more available frequency bands, for instance, the use of system different parameters can provide more frequency bands as shown in Figs. 3.14-3.16. The generated carrier signals can be used as the modulated carrier that can be used to form the simultaneous up and down link and multi switching, which is controlled by a computer server. This can also be used with the existed public network installation or Ad Hoc network applications. Furthermore, the pumping part is not required in such a system. The new available frequency bands can be use to form the new multi-frequency layer protocol, where more communication capacity can be performed.

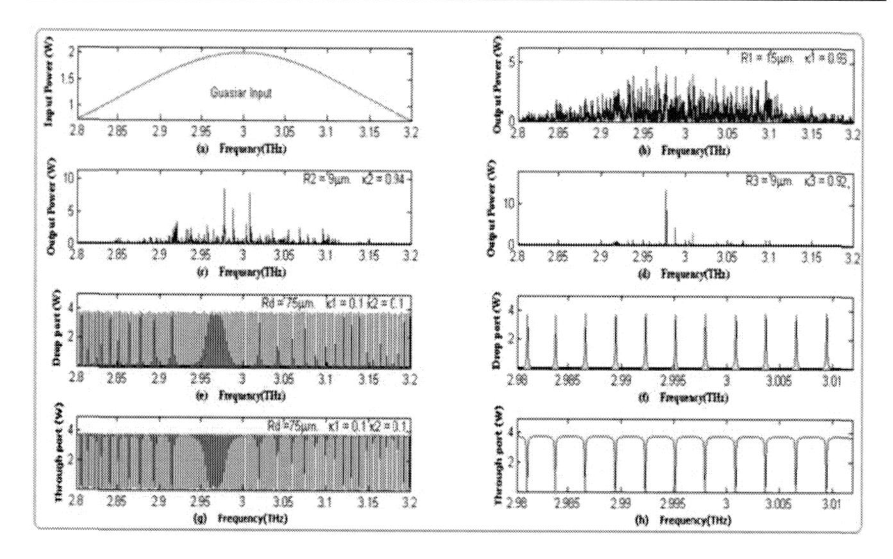

Figure 3.14. Results of the THz frequency band with the center frequency at 3THz, where (a) the input Gaussian pulse, (b) the large bandwidth signal, (c) the filtering and amplifying signals, (d) output frequency band, (e) and (f)are the drop port signals, (g) and (h)are the through port signals.

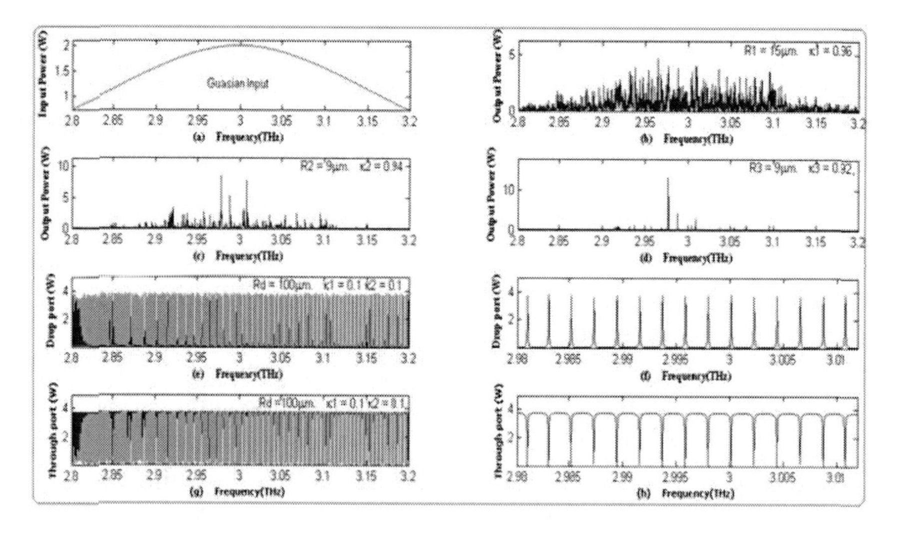

Figure 3.15. Results of the THz frequency band with the center frequency at 3THz, where (a) the input Gaussian pulse, (b) the large bandwidth signal, (c) the filtering and amplifying signals, (d) output frequency band, (e) and (f)are the drop port signals, (g) and (h)are the through port signals.

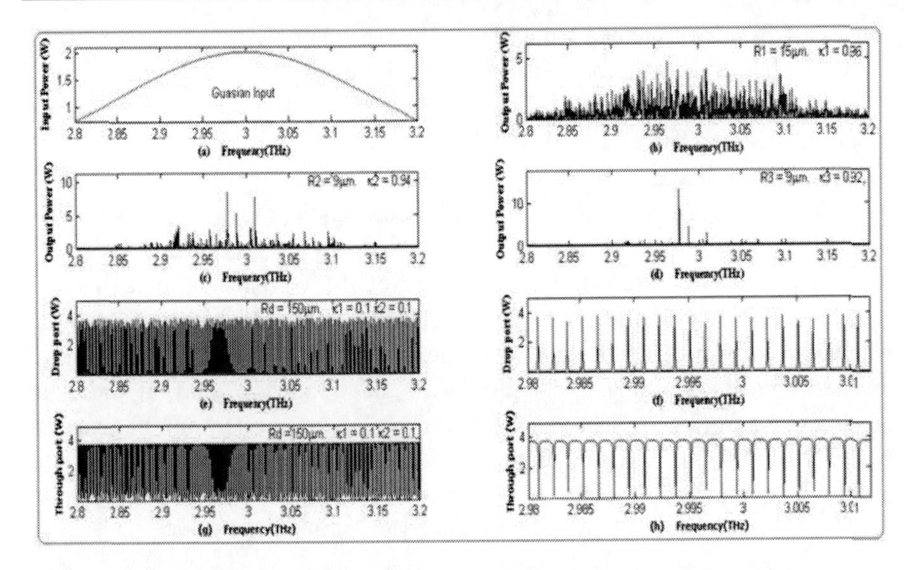

Figure 3.16. Results of the THz frequency band with the center frequency at 3THz, where (a) the input Gaussian pulse, (b) the large bandwidth signal, (c) the filtering and amplifying signals, (d) output frequency band, (e) and (f)are the drop port signals, (g) and (h)are the through port signals.

3.5. CONCLUSION

We have shown that the multi-wavelength bands can be generated by using a Gaussian pulse propagating within the microring resonator system, which is available for the extended DWDM with the wavelength center at 0.40-1.50 μm, which can be used with the existed public networks, where the non-dispersive wavelength (1.30 μm) can be extended and used to increase the communication capacity, furthermore, for long distance link, the pumping is not required in such a system. One of the results with the center wavelength at 1.50 μm has shown that the spatial pulse width of 1.0 nm and the spectrum range of 3.0 nm can be generated and achieved, which is as shown in Figure 3.7(b). Moreover, the problem of signal collision can be solved by using the suitable FSR design [19]. In general, by using the wider range of ring parameters, the spectral range of the output can be covered wider range instead of fraction of nm. The large increasing in peak power is seen when light propagates from the large to small effective core area, where the other parameter is the coupling coefficient. However, the amplified power is required to control to keep the device being realistic.

Apart from communication application, the idea of personnel wavelength (network) being realistic for the large demand user due to un-limit wavelength discrepancy, whereas the specific soliton band can be generated using the proposed system. The potential of soliton bands such as visible soliton (color soliton), UV-soliton, X-ray soliton and infrared soliton can be generated and used for the applications such as multi color holography, medical tools, security imaging and transparent holography and detection, respectively.

Finally, We have shown that the multi frequency bands can be generated by using a Gaussian pulse propagating within the microring resonator system, which can be simultaneous link within a single device and available for the extended multi switching application with the frequency relay at the THz band. The Mirroring resonators system embedded in mobile node to generate bandwidth for serve two communication styles are direct communication and multi-hop communication by relay service. This can be used for wireless network with the existed public networks or the Ad hoc network.

REFERENCES

[1] D. Deng and Q. Guo, "Ince-Gaussian solitons in strongly nonlocal nonlinear media", *Opt. Lett.*, 32(2007)3206-3208.

[2] G. Xia, Z. Wu, and J. Wu, "Effect of fiber chromatic dispersion on incident super-Gaussian pulse transmission in single-mode fibers," *Chinese J. Phys.*, 41(2)(2003)118-120.

[3] S. Supparpola, Y. Sun and S. A. Chiramida, "Gaussian pulse decomposition: An intuitive model of electrocardiogram waveforms," *Annals of Biomedical Engineering*, 25(1997)252-260.

[4] P.K.A. Wai, K. Nakkeeran, "On the uniqueness of Gaussian ansatz parameters equations: generalized projection operator method," *Phys. Lett., A*, 332 (2004)239–243.

[5] P.P. Yupapin and W. Suwancharoen, "Chaotic signal generation and cancellation using a microring resonator incorporating an optical add/drop multiplexer, *Opt. Commun.*, 280(2007)343-350.

[6] P.P. Yupapin, N. Pornsuwanchroen and S. Chaiyasoonthorn, "Attosecond pulse generation using nonlinear microring resonators," *Microw. and Opt. Technol. Lett.,* 50(2008)3108-3111.

[7] N. Pornsuwancharoen and P.P. Yupapin, "Generalized fast, slow, stop, and store light optically within a nanoring resonator," *Microw. and Opt. Technol. Lett.,* 51(2009) 899-902.

[8] Y. S. Kivshar and B. Luther-Davies, "Dark optical solitons: Physics and applications," *Phys. Rep.*, 298(1998)81-197.

[9] A. Hasegawa (Editor), "*Massive WDM and TDM Soliton Transmission Systems*," Kluwer Academic Publishers, Boston, 2000.

[10] G. P. Agrawal, "*Nonlinear Fiber Optics*", Academic Press, New York, 1995.

[11] Yu. A. Simonov and J.A. Tjon, "Soliton-soliton interaction in confining models," *Phys. Lett., B* 85(1979)380-384.

[12] J.K. Drohm, L.P. Kok, Yu. A. Simonov, J.A. Tjon and A.I. Veselov, "Collision and rotation of solitons in three space-time dimensions," *Phys. Lett.,* B101(1981)204-208.

[13] T. Iizuka and Y. S. Kivshar "Optical gap solitons in nonresonant quadratic media, *Phys. Rev. E* 59(1999)7148 - 7151.

[14] R. Ganapathy, K. Porsezian, A. Hasegawa, V.N. Serkin, "Soliton interaction under soliton dispersion management," *IEEE Quantum Electron*, 44(2008)383-390.

[15] N. Pornsuwancharoen, U. Dunmeekaew and P.P. Yupapin, "Multi-soliton generation using a micro ring resonator system for DWDM based soliton communication," *Microw. and Opt. Technol. Lett.*, 51(5), 1374-1377(2009).

[16] P.P. Yupapin, N. Pornsuwanchroen and S. Chaiyasoonthorn, "Attosecond pulse generation using nonlinear micro ring resonators," *Microw. and Opt. Technol. Lett.*, 50(12), 3108-3011(2008).

[17] N. Pornsuwancharoen and P.P. Yupapin, "Generalized fast, slow, stop, and store light optically within a nano ring resonator," *Microw. and Opt. Technol. Lett.*, 51(4), 899-902(2009).

[18] N. Pornsuwancharoen, S. Chaiyasoonthorn and P.P. Yupapin, "Fast and slow lights generation using chaotic signals in the nonlinear micro ring resonators for communication security," *Opt. Eng.*, 48(1), 50005-1-5(2009).

[19] P.P. Yupapin, P. Saeung and C. Li, "Characteristics of complementary ring-resonator add/drop filters modeling by using graphical approach," *Opt. Commun.*, 272, 81-86(2007).

[20] M. Fujii, J. Leuthold and W. Freude, "Dispersion relation and loss of subwavelength confined mode of metal-dielectri-gap optical waveguides, " *IEEE Photon. Technol. Lett.,* 21, 362-364(2009).

[21] P.P. Yupapin and N. Pornsuwancharoen, "Proposed nonlinear micro ring resonator arrangement for stopping and storing light," *IEEE Photon. Technol. Lett.*, 21, 404-406(2009).

[22] N. Ryu, Y. Yun, S. Choi,R. C. Palat, AND J. H. Reed, “Smart antenna base station open architecture for SDR networks” *IEEE J. wireless comm.,* 13(3)(2006)322-336.

[23] Z. Wang, H. R. Sadjadpour,J. J. Garcia-Luna-Aceves, S. S. Karande, “Fundamental limits of information dissemination in wireless Ad Hoc networks—part I: single-packet reception”, *IEEE Trans. on Wireless comm.*,8(12)(2009)5749-5754.

[24] C. L. Hu, M. S. Chen, “Adaptive information dissemination: An extended wireless data broadcasting scheme with loan-based feedback control”, *IEEE Trans. on Mobile Computing*, 2(4)(2003)322–336.

[25] S. Nittel, M. Duckham, L. Kulik, “Information dissemination in mobile Ad Hoc geosensor networks”, *Lecture Notes in Computer Science* 3234, Springer,(2004)206-222.

[26] A. Nedos, K. Singh, R. Cunningham,S. Clarke, “Probabilistic discovery of demantically diverse content in MANETs”, *IEEE Trans. on mobile Comp.,* 8(4)(2009)544-557.

[27] H. Gharavi, “Multichannel mobile Ad Hoc links for multimedia communications”, *IEEE Proc.,* 96(1)(2008)77-95.

[28] P. Mohapatra, C. Gui, and J. Li, "Group communications in mobile Ad Hoc networks", *IEEE Spec. Ad Hoc Networks*, (2004)70-77.

[29] C. C. Shen, C. Srisathapornphat, R. Liu, Z. Huang, C. Jaikaeo, S. and E. L. Lloyd, "CLTC: A cluster-based topology control framework for Ad Hoc networks", *IEEE Trans. Mob. Comp*.,3(1)(2004)18-32.

[30] S. G. WANG, H. JI, T. LI, and J. G. MEI, "Topology-aware peer-to-peer overlay network for Ad-hoc", *J. Posts and Tele*.,16(1)(2009)111-115.

[31] A. Zemlianov, G. D. Veciana, " Capacity of Ad hoc wireless networks with infrastructure support", *IEEE J. Comm*.,23(3)(2005)657-667.

[32] C. Comaniciu, H. V. Poor, "On the capacity of mobile Ad hoc networks with delay constraints", *IEEE Trans. Wirel. Comm*., 5(8)(2006)2061-2071.

[33] G. Jakllari, S. V. Krishnamurthy, M. Faloutsos, and P. V. Krishnamurthy, "On broadcasting with cooperative diversity in Multi-Hop wireless networks ", *IEEE J. Comm*., 25(2)(2007)484-495.

[34] A. Sharma, G. Singh, “Rectangular microstirp patch antenna design at THz frequency for short distance wireless communication systems”, *J. Infrared Milli Terahz Waves*, 30(2009)1-7.

[35] Q. Xu and M. Lipson, “All-optical logic based on silicon micro-ring resonators,” *Opt. Express* 15(3)(2007)924-929.

[36] Y. Su, F. Liu and Q. Li, "System performance of slow-light buffering and storage in silicon nano-waveguide," *Proc. SPIE*, 6783, 68732P(2007).

[37] Y. Kokubun, Y. Hatakeyama, M. Ogata, S. Suzuki, and N. Zaizen, "Fabrication technologies for vertically coupled micro ring resonator with multilevel crossing busline and ultracompact-ring radius", *IEEE J. Sel. Top. Quantum Electron.,* 11(2005)4–10.

[38] S. Mithata, N. Pornsuwancharoen and P.P. Yupapin, "A Simultaneous short wave and millimeter wave generation using a soliton pulse within a nano-waveguide" , *IEEE Photon. Technol. Lett.*, 21(13)(2009)932-934.

Chapter 4

RANDOM BINARY CODE GENERATION

4.1. INTRODUCTION

All optical devices are becoming important equipments for advanced optical technology due to the increasing demand for processing speed in various applications such as multimedia and telemedicine has attracted significant interest in developing all-optical signal processing technologies for future photonic networks. In operation, optical signal processing functions [1], including pulse repetition rate multiplication (PRRM) and arbitrary waveform generation (AWG), are useful to generate ultrafast pulse trains with binary amplitude code patterns for ultra-wideband applications or as codes for labels in label-switched networks and optical code-division multiple access. There are evidences of growing interest in either theory or experiment of photonic integrated circuits (PICs) for high-speed all-optical signal processing development. Xia et al [2] have demonstrated that the 10-40 GHz pulse repetition rate multiplication with amplitude control using a 4-stage lattice-form Mach-Zehnder interferometer (LF-MZI) was fabricated in silica-on-silicon planar lightwave circuit (PLC) technology. Jain et al [3] have reported the design and analysis of a 6-stage tunable LF-MZI for arbitrary binary code generation at 40 GHz by used the direct temporal domain approach to determine the filter parameters which could be tuned via the thermo-optic effect in silica; using numerical simulations and demonstrated that the 6-stage device could generate any 4-bit binary amplitude or phase pattern at 40 GHz from a uniform 10 GHz input pulse train. Ou et al [4] have demonstrated a novel optical generation approach for binary phase-coded, direct sequence ultra-wideband (UWB) signals. A system consists of a laser array, a polarization modulator (PolM), a fiber Bragg grating (FBG), a length of single mode fiber, and a photo detector (PD). The FBG, designed based on the superimposed, chirped grating, is used as the multi-channel frequency discriminator. The input electronic Gaussian pulse is modulated on the optical carrier by the PolM and then converted into UWB monocycle or doublet pulses sequence by the multi-channel frequency discriminator. The PolM is used so that the desired binary phase code pattern could be simply selected by adjusting the polarization state of each laser, rather than tuning the laser wavelengths. The desired UWB shape, monocycle or doublet, could be selected by tuning the FBG. Based on our proposed approach, four-chip, binary-phase-coded. An optical coding scheme using optical interconnection for a photonic analog-to-digital conversion has been proposed and demonstrated [5]. It allows us to convert a multi-power level signal into a multiple-bit binary code so as to detect it in a bit-parallel format by binary photodiode array. The proposed optical coding is executed after optical quantization using self-frequency shift. Optical interconnection based on a binary conversion table generates a multiple-bit binary code by appropriate allocation of a level identification signal which is provided as a result of optical quantization.

Recently, an experimentally demonstration of a novel technique to implement bipolar UWB pulse coding has been reported[6], in which a multichannel chirped FBG was used, in combination with a dispersive fiber, to produce a multichannel frequency discriminator with a step-increased group-delay response. Binary monocycle or doublet sequences with the desired phase coding patterns were generated by applying a phase modulated Gaussian pulse train to the multichannel frequency discriminator. The time delay difference between adjacent UWB pulses remains unchanged when the wavelengths are tuned. Different phase coding patterns were generated by simply tuning the states of polarization of the wavelengths sent to the PolM. In both cases, the dispersion of the dispersive fiber was compensated by the multichannel chirped FBG; therefore the distortion of the generated UWB pulses due to the fiber dispersion was eliminated. A novel and simple method for all-optically generating UWB pulses using a nonlinear optical loop mirror (NOLM) based optical switch. Both Gaussian monocycle and doublet pulses and their polarity reversed pulses have been proposed [7], whose spectra accord with Federal Communications Commission (FCC) standard, are generated experimentally and distributed by the SMF, which acting as both dispersive and transmission media. A tunable laser adopted in the scheme ensures that the different length of SMF has no effects on the generated UWB pulses. Furthermore, pulse shape modulation at a high speed can easily achieved by employing an intensity modulator. The authors in reference [8] have reported a Barker binary phase code for speckle reduction in line scan laser projectors, and a speckle contrast factor decrease down to 13%. Barker-like binary phase codes of lengths longer than 13 are used at an intermediate image plane. It is shown by theoretical calculation that a much better speckle reduction with a speckle contrast factor up to 6% can be achieved by using longer binary phase codes other than the Barker code.

From the previous reports, the searching of new optical encoding technique remains, therefore, in this chapter we propose the use of dark-bright optical soliton conversion and control within a tiny ring resonator system [9, 10]. It is formed by using an all optical devices, which consists of an add/drop optical filter and known as a PANDA ring resonator [11], which is a proposed model of ring resonator. The binary codes can be formed and retrieved by using the random polarized light and dark-bright soliton conversion, respectively. The advantage of this technique is that the random binary codes can be formed by controlling one of the input solitons into the PANDA ring resonator ports, which can be available for the use in information security requirement, moreover, the device dimension can be used to form an array/large photonic circuits, whereas the coding capacity can be increased.

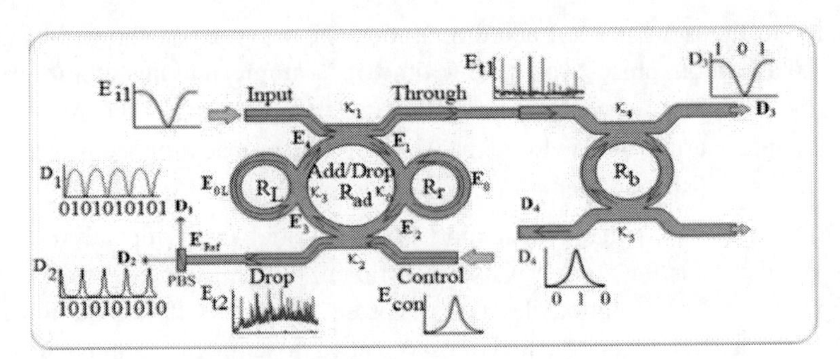

Figure 4.1. Schematic diagram of random binary code generation using dark-bright soliton conversion. D_n: photodetectors, κ_n: coupling coefficient of couplers, E_{Ref}: references filed, E_n : electric fields, PBS: Polarized Beam Splitter.

4.2. Theoretical Modelling

The proposed system consists of an add/drop filter and double nanoring resonators as shown in figure 1. To perform the dark-bright soliton conversion, dark and bright solitons are input into the add/drop optical filter system, whereas the input optical field (E_{i1}) and the control port optical field (E_{con}) of the dark-bright solitons pulses are given by [12]

$$E_{i1}(t) = A\tanh\left[\frac{T}{T_0}\right]\exp\left[\left(\frac{x}{2L_D}\right) - i\phi(t)\right], \tag{4.1a}$$

$$E_{con}(t) = A\,\mathrm{sech}\left[\frac{T}{T_0}\right]\exp\left[\left(\frac{x}{2L_D}\right) - i\phi(t)\right], \tag{4.1b}$$

in which A and z are the optical field amplitude and propagation distance, respectively. $\phi(t) = \phi_0 + \phi_{NL} = \phi_0 + \frac{2\pi n_2 L}{A_{eff}\lambda}|E_0(t)|^2$ is the random phase term related to the temporal coherence function of the input light, ϕ_0 is the linear phase shift, ϕ_{NL} is the nonlinear phase shift, n_2 is the nonlinear refractive index of InGaAsP/InP waveguide. The effective mode core area of the device is given by A_{eff}., $L = 2\pi R_{ad}$, R_{ad} is the radius of device, λ is the input wavelength light field and $E_0(t)$ is the circulated field within nanoring coupled to the right and left add/drop optical filter system as shown in figure 1. T is a soliton pulse propagation time in a frame moving at the group velocity, $T=t-\beta_1 z$, where β_1 and β_2 are the coefficients of the linear and second-order terms of Taylor expansion of the propagation constant. $L_D= T_0^2/|\beta_2|$ is the dispersion length of the soliton pulse. T_0 in Equation (4.1a) and (4.1b) is a soliton pulse propagation time at initial input (or soliton pulse width), where t is the soliton phase shift time, and the frequency shift of the soliton is ω_0. This solution describes a pulse that keeps its temporal width invariance as it propagates, and thus is called a temporal soliton. When a soliton of peak intensity $(|\beta_2/\Gamma T_0^2|)$ is given, then T_0 is known. For the soliton pulse in the microring device, a balance should be achieved between the dispersion length (L_D) and the nonlinear length ($L_{NL}=1/\Gamma\phi_{NL}$), where $\Gamma=n_2 k_n$, is the length scale over which dispersive or nonlinear effects makes the beam become wider or narrower. For a soliton pulse, hence $L_D = L_{NL}$.

When light propagates within a nonlinear medium, the refractive index (n) of light within the medium is given by

$$n = n_0 + n_2 I = n_0 + \frac{n_2}{A_{eff}} P, \tag{4.2}$$

with n_0 and n_2 as the linear and nonlinear refractive indexes, respectively. I and P are the optical intensity and the power, respectively. The effective mode core area of the device is given by A_{eff}. For the add/drop optical filter design, the effective mode core areas range from 0.50 to 0.10 μm^2, in which the parameters were obtained by using the related practical material parameters (InGaAsP/InP)[13-15]. When a dark soliton pulse is input and propagated within a add/drop optical filter as shown in Figure 4.1, the resonant output is formed

In Figure 4.1, consists of an add/drop optical multiplexer used for generated random binary coded light pulse and an add/drop optical filter device for decoded binary code signal. The resonator output field, E_{t1} and E_1 consists of the transmitted and circulated components within the add/drop optical multiplexing system, which provides the driving force to photon/molecule/atom.

When the input light pulse passes through the first coupler of the add/drop optical multiplexing system, the transmitted and circulated components can be written as

$$E_{t1} = \sqrt{1-\gamma_1}\left[\sqrt{1-\kappa_1}E_{i1} + j\sqrt{\kappa_1}E_4\right] \tag{4.3}$$

$$E_1 = \sqrt{1-\gamma_1}\left[\sqrt{1-\kappa_1}E_4 + j\sqrt{\kappa_1}E_{i1}\right] \tag{4.4}$$

$$E_2 = E_0 E_1 e^{-\frac{\alpha}{2}\frac{L}{2} - jk_n \frac{L}{2}} \tag{4.5}$$

where κ_1 is the intensity coupling coefficient, γ_1 is the fractional coupler intensity loss, α is the attenuation coefficient, $k_n = 2\pi/\lambda$ is the wave propagation number, λ is the input wavelength light field and $L = 2\pi R_{ad}$, R_{ad} is the radius of add/drop device.

For the second coupler of the add/drop optical multiplexing system,

$$E_{t2} = \sqrt{1-\gamma_2}\left[\sqrt{1-\kappa_2}E_{i2} + j\sqrt{\kappa_2}E_2\right] \tag{4.6}$$

$$E_3 = \sqrt{1-\gamma_2}\left[\sqrt{1-\kappa_2}E_2 + j\sqrt{\kappa_2}E_{i2}\right] \tag{4.7}$$

$$E_4 = E_{0L}E_3 e^{-\frac{\alpha}{2}\frac{L}{2} - jk_n\frac{L}{2}} \tag{4.8}$$

Here κ_2 is the intensity coupling coefficient, γ_2 is the fractional coupler intensity loss. The circulated light fields, E_0 and E_{0L} are the light field circulated components of the nanoring radii, R_r and R_L which couples into the right and left sides of the add/drop optical multiplexing system, respectively. The light field transmitted and circulated components in the right nanoring, R_r, are given by

$$E_2 = \sqrt{1-\gamma}\left[\sqrt{1-\kappa_0}E_1 + j\sqrt{\kappa_0}E_{r2}\right] \tag{4.9}$$

$$E_{r1} = \sqrt{1-\gamma}\left[\sqrt{1-\kappa_0}E_{r2} + j\sqrt{\kappa_0}E_1\right] \tag{4.10}$$

$$E_{r2} = E_{r1}e^{-\frac{\alpha}{2}L_1 - jk_nL_1} \tag{4.11}$$

where κ_0 is the intensity coupling coefficient, γ is the fractional coupler intensity loss, α is the attenuation coefficient, $k_n = 2\pi/\lambda$ is the wave propagation number, λ is the input wavelength light field and $L_1 = 2\pi R_r$, R_r is the radius of right nanoring.

From Eqs. (4.9)-(4.11), the circulated roundtrip light fields of the right nanoring radii, R_r, are given in Eqs. (4.12) and (4.13), respectively.

$$E_{r1} = \frac{j\sqrt{1-\gamma}\sqrt{\kappa_0}E_1}{1-\sqrt{1-\gamma}\sqrt{1-\kappa_0}e^{-\frac{\alpha}{2}L_1 - jk_nL_1}} \tag{4.12}$$

$$E_{r2} = \frac{j\sqrt{1-\gamma}\sqrt{\kappa_0}E_1 e^{-\frac{\alpha}{2}L_1 - jk_nL_1}}{1-\sqrt{1-\gamma}\sqrt{1-\kappa_0}e^{-\frac{\alpha}{2}L_1 - jk_nL_1}} \tag{4.13}$$

Thus, the output circulated light field, E_0, for the right nanoring is given by

$$E_0 = E_1 \left\{ \frac{\sqrt{(1-\gamma)(1-\kappa_0)} - (1-\gamma)e^{-\frac{\alpha}{2}L_1 - jk_n L_1}}{1-\sqrt{(1-\gamma)(1-\kappa_0)}e^{-\frac{\alpha}{2}L_1 - jk_n L_1}} \right\} \quad (4.14)$$

Similarly, the output circulated light field, E_{0L}, for the left nanoring at the left side of the add/drop optical multiplexing system is given by

$$E_{0L} = E_3 \left\{ \frac{\sqrt{(1-\gamma_3)(1-\kappa_3)} - (1-\gamma_3)e^{-\frac{\alpha}{2}L_2 - jk_n L_2}}{1-\sqrt{(1-\gamma_3)(1-\kappa_3)}e^{-\frac{\alpha}{2}L_2 - jk_n L_2}} \right\} \quad (4.15)$$

where κ_3 is the intensity coupling coefficient, γ_3 is the fractional coupler intensity loss, α is the attenuation coefficient, $k_n = 2\pi/\lambda$ is the wave propagation number, λ is the input wavelength light field and $L_2 = 2\pi R_L$, R_L is the radius of left nanoring.

From Eqs. (4.3)-(4.15), the circulated light fields, E_1, E_3 and E_4 are defined by given $x_1 = (1-\gamma_1)^{1/2}$, $x_2 = (1-\gamma_2)^{1/2}$, $y_1 = (1-\kappa_1)^{1/2}$, and $y_2 = (1-\kappa_2)^{1/2}$.

$$E_1 = \frac{jx_1\sqrt{\kappa_1}E_{i1} + jx_1x_2y_1\sqrt{\kappa_2}E_{0L}E_{i2}e^{-\frac{\alpha}{2}\frac{L}{2} - jk_n\frac{L}{2}}}{1 - x_1x_2y_1y_2E_0E_{0L}e^{-\frac{\alpha}{2}L - jk_n L}} \quad (4.16)$$

$$E_3 = x_2y_2E_0E_1e^{-\frac{\alpha}{2}\frac{L}{2} - jk_n\frac{L}{2}} + jx_2\sqrt{\kappa_2}E_{i2} \quad (4.17)$$

$$E_4 = x_2y_2E_0E_{0L}E_1e^{-\frac{\alpha}{2}L - jk_n L} + jx_2\sqrt{\kappa_2}E_{0L}E_{i2}e^{-\frac{\alpha}{2}\frac{L}{2} - jk_n\frac{L}{2}} \quad (4.18)$$

Thus, from Eqs. (4.3), (4.5), (4.16)-(4.18), the output optical field of the through port (E_{t1}) is expressed by

$$E_{t1} = x_1 y_1 E_{i1} + \begin{pmatrix} jx_1 x_2 y_2 \sqrt{\kappa_1} E_0 E_{0L} E_1 \\ -x_1 x_2 \sqrt{\kappa_1 \kappa_2} E_{0L} E_{i2} \end{pmatrix} e^{-\frac{\alpha}{2}\frac{L}{2} - jk_n \frac{L}{2}} \tag{4.19}$$

The power output of the through port (P_{t1}) is written by

$$P_{t1} = (E_{t1}) \cdot (E_{t1})^* = |E_{t1}|^2 . \tag{4.20}$$

Similarly, from Eqs. (4.5), (4.6), (4.16)-(4.18), the output optical field of the drop port (E_{t2}) is given by

$$E_{t2} = x_2 y_2 E_{i2} + jx_2 \sqrt{\kappa_2} E_0 E_1 e^{-\frac{\alpha}{2}\frac{L}{2} - jk_n \frac{L}{2}} \tag{4.21}$$

The power output of the drop port (P_{t2}) is expressed by

$$P_{t2} = (E_{t2}) \cdot (E_{t2})^* = |E_{t2}|^2 . \tag{4.22}$$

In order to retrieve the required signals, we propose to use the add/drop optical multiplexing device with the appropriate parameters. This is given in the following details. The optical circuits of ring resonator add/drop multiplexing for the through port and drop port can be given by Eqs. (4.20) and (4.22), respectively. The chaotic noise cancellation can be managed by using the specific parameters of the add/drop multiplexing device, and the required signals can be retrieved by the specific users. κ_1 and κ_2 are the coupling coefficients of the add/drop filters, $k_n=2\pi/\lambda$ is the wave propagation number for in a vacuum, and the waveguide (ring resonator) loss is α = 5×10^{-5} dBmm^{-1}. The fractional coupler intensity loss is γ = 0.01. In the case of the add/drop multiplexing device, the nonlinear refractive index is neglected.

The electric field detected by photodectector D_3 is given by [16];

$$E_{D_3} = E_{t1} \frac{-\sqrt{1-\kappa_4} e^{-\frac{\alpha}{2} L_b - jk_n L_b} + \sqrt{1-\kappa_4}}{1 - \sqrt{1-\kappa_4} \sqrt{1-\kappa_5} e^{-\frac{\alpha}{2} L_b - jk_n L_b}} \tag{4.23}$$

where $L_b = 2\pi R_b$, R_b is radius of add/drop optical filter decoded as shown in figure 1. The light pulse output power detected by photodectector D_3 is defined as

$$P_{D_3} = \left|\frac{E_{D_3}}{E_{t1}}\right|^2 = \frac{\begin{pmatrix} 1-\kappa_4 - 2\sqrt{1-\kappa_4}\sqrt{1-\kappa_5}e^{-\frac{\alpha}{2}L_b} \\ \times\cos(k_n L_b) + (1-\kappa_4)e^{-\alpha L_b} \end{pmatrix}}{\begin{pmatrix} 1+(1-\kappa_4)(1-\kappa_5)e^{-\alpha L_b} \\ -2\sqrt{1-\kappa_4}\sqrt{1-\kappa_5}e^{-\frac{\alpha}{2}L_b}\cos(k_n L_b) \end{pmatrix}}. \quad (4.24)$$

The electric field detected by photodectector D_4 is given by

$$E_{D_4} = E_{t1}\frac{-\sqrt{\kappa_4\kappa_5}e^{-\frac{\alpha}{2}\frac{L_b}{2} - jk_n\frac{L_b}{2}}}{1-\sqrt{1-\kappa_4}\sqrt{1-\kappa_5}e^{-\frac{\alpha}{2}L_b - jk_n L_b}} \quad (4.25)$$

The light pulse output power detected by photodectector D_4 is defined as

$$P_{D_4} = \left|\frac{E_{D_4}}{E_{t1}}\right|^2 = \frac{\kappa_4\kappa_5 e^{-\frac{\alpha}{2}L_b}}{\begin{pmatrix} 1+(1-\kappa_4)(1-\kappa_5)e^{-\alpha L_b} \\ -2\sqrt{1-\kappa_4}\sqrt{1-\kappa_5}e^{-\frac{\alpha}{2}L_b}\cos(k_n L_b) \end{pmatrix}}. \quad (4.26)$$

For the random binary code generation, we used the add/drop optical filter for decoded binary code signal as shown in Figure 4.1. When the light pulse signal passes through the coupler with coupling coefficient, $\kappa_4 = 0.5$, light will be split into two ways, once passes through photodectector, D_3 and detected output binary code signal and other circulated within the add/drop optical filter and passes through the coupler which coupling coefficient, $\kappa_5 = 0.5$, to photodetector, D_4 and detected once output binary code signal.

4.3. Random Binary Code Generation

In simulation, the used parameters of the first add/drop filter (optical multiplexer, PANDA ring resonator) are fixed to be $\kappa_0 = 0.1$, $\kappa_1 = 0.35$, $\kappa_2 =$

0.1, and κ_3 = *0.2* respectively. The ring radii are R_{ad} = *300nm,* R_r = *30nm,* and R_l = *15nm*. A_{eff} are *0.50, 0.25* and *0.25* μm^2 [13] for the add/drop optical multiplexer, right and left nanoring resonators, respectively. The parameters of the second add/drop optical filter are fixed to be κ_4 = κ_5 = *0.5*, R_b = *200nm* and A_{eff} = *0.25* μm^2, respectively. Simulation results of the random binary code generation with center wavelengths are at λ_0 = *1.50*μm. For random binary code generation is as shown in Figure 4.2, the simulation result of the light pulse generated within the add/drop optical filter system at center wavelength λ_0 = *1.50* μm for random binary code generation, where (a) $|E_1|^2$, (b) $|E_2|^2$, (c) $|E_3|^2$, (d) $|E_4|^2$ and (e) are the reflected outputs from the throughput port, and (f) is the output at the drop port. Here (a) E_1, (b) E_2, (c) E_3 and (d) E_4 (e) are the through port and (f) drop port signals. When a dark soliton light pulse with 1W peak power is input into the input port traveled and passes through the first coupler, κ_1, which one part split into to through port and other to arc at E_1 of add/drop optical multiplexer. The cross phase modulation (XPM) output power is as shown in Figure 4.2(a). Then the light pulse enters and circulates in the right nanoring resonator radii, R_r, which it passes to the add/drop optical multiplexer at E_2, The output power has amplified (larger amplitude) as shown in Figure 4.2(b).

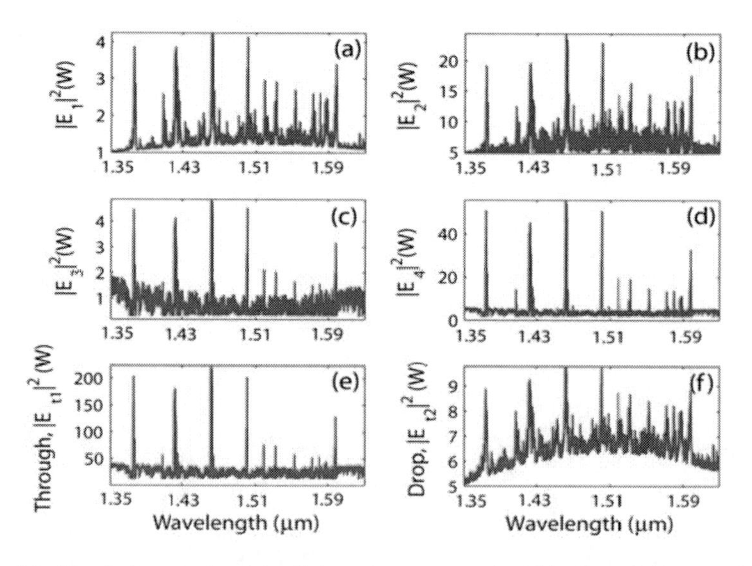

Figure 4.2. Simulation result of the light pulse generated within the add/drop optical filter system at center wavelength λ_0 = *1.50* μm for random binary code generation, where (a) $|E_1|^2$, (b) $|E_2|^2$, (c) $|E_3|^2$, (d) $|E_4|^2$, (e) are through port and (f) drop port signals, where R_r = *15nm,* R_L = *30nm,* R_{ad} = *300nm* and $\alpha = 5 \times 10^{-5}$ $dBmm^{-1}$.

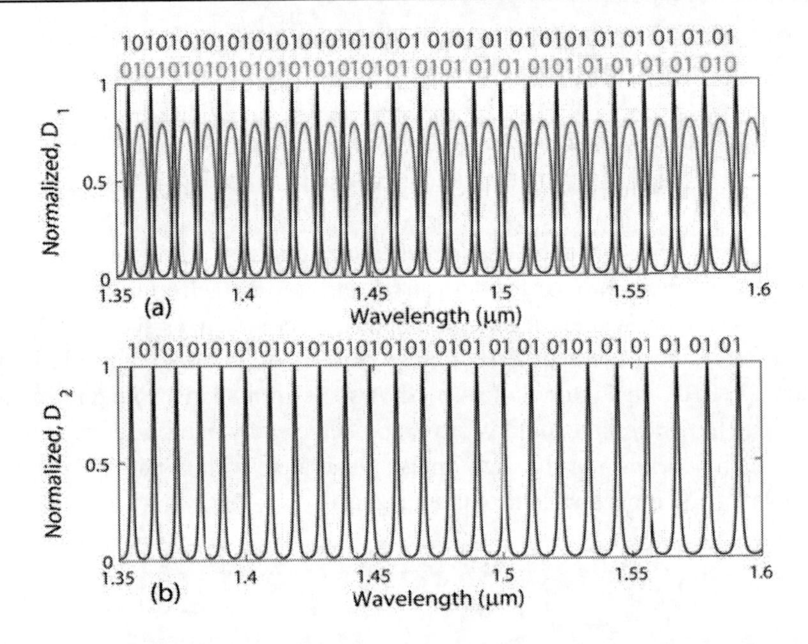

Figure 4.3. Random binary code generation for reference output signal, E_{Ref} , passes through PBS at the drop port, which (a) detected by D_1 and (b) detected by D_2.

To use the control function, a bright soliton with 1W peak power is input into the control port, then it passes through the second coupler, κ_2, and it is multiplexed with the light pulse from E_2. The output light is then split into two ways. One part is split into drop port, Et_2, the power output as shown in Figure 4.2(e), and other is split into the arc of add/drop at position E_3 with the output power as shown in Figure 4.2(c). Finally, the light pulse travels and enters into the left nanoring resonator radii, R_L, and then passes to the add/drop optical multiplexer at position E_4. The output power is amplified again as shown in Figure 4.2(d) and travels into the first coupler, κ_1. It then enters into the through port, Et_1, and E_1, whereas the amplitude is further amplified that the one as shown in Figure 4.2(f).

In operation, the random binary codes are generated within the add/drop optical multiplexer, which are seen at the drop port (Figure 4.2 (f)). It travels through the PBS, where the polarization phase shift of thetwo components is 90°, which means that random polarization states of two components can be used to form the random binary code patterns and the binary code signals. These can be observed at photodetectors D_1 and D_2. The referencing binary code patterns are set as shown in Figure 4.3(a) and 4.3(b), where the binary codes detected by D_1 are

'1010101010101010101010101010101010101010101010'and
'101010101010101010101010101010101 0101010101010' patterns.

The binary codes detected by D_2 are
'1010101010101010101010101010101010101010101010' patterns.

In this work, the obtained pulse switching time of 12.8ns is noted. In operation, the random binary codes can be retrieved in the form of dark-bright soliton conversion within the add/drop optical filter system as shown in Figure 4.4, whereas the pattern of dark soliton conversion is '101', and the bright soliton is '010'. In Figure 4.4(a)-(b) shows the formation of random binary codes due to dark-bright soliton conversion detected by photodetectors D_3 and D_4, which depends on the light pulse generated within add/drop optical multiplexer and input and control input signals.

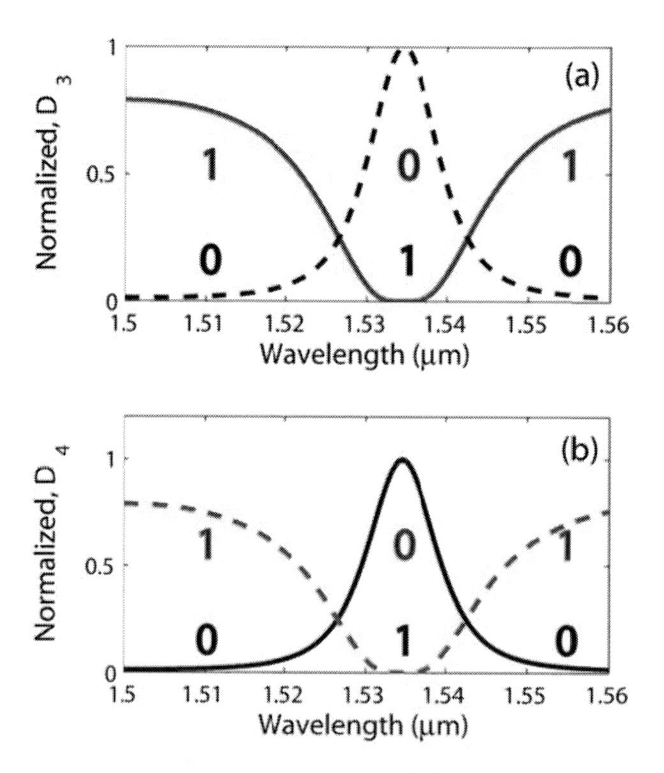

Figure 4.4. Binary code signal form dark-bright solitons conversion, where (a) dark and (b) bright, and $\kappa_4 = \kappa_5 = 0.5$, $R_b = 200nm$.

4.4. Random Code and Security

The proposed system consists of a suppression system and recovery system as shown in Figures 4.5 and 4.6. The binary code suppression is designed by using the add/drop filter and double microring resonators, which it is known as PANDA ring resonator. The binary code recovery is obtained by using add/drop filter. In simulation, the parameters of PANDA ring resonator are fixed to be $\kappa_0 = 0.1$, $\kappa_1 = 0.2$, $\kappa_2 = 0.2$, and $\kappa_3 = 0.1$ respectively. The ring radii are $R_{ad} = 200\mu m$, $R_r = 15\mu m$, and $R_l = 15\mu m$. A_{eff} are 0.50, 0.25 and 0.25 μm^2 [13, 15, 17] for PANDA ring resonator, right and left microring resonators, respectively. For the binary code recovery, the parameters of the add/drop optical filter are fixed to be $\kappa_4 = \kappa_5 = 0.2$, $R_b = 50\mu m$ and $A_{eff} = 0.25$ μm^2, respectively. Moreover, our binary suppression and recovery system is complaint to the possible fabricated devices. Simulation result of the binary code suppression with center wavelength is at $\lambda_0 = 1.50\mu m$ is presented. The binary code suppression result is as shown in Figure 4.7. The binary code recovery result is as shown in Figure 4.8.

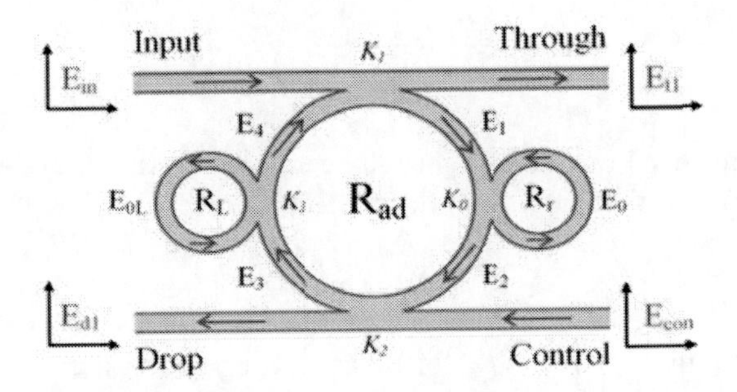

Figure 4.5. Schematic diagram of binary code suppression system.

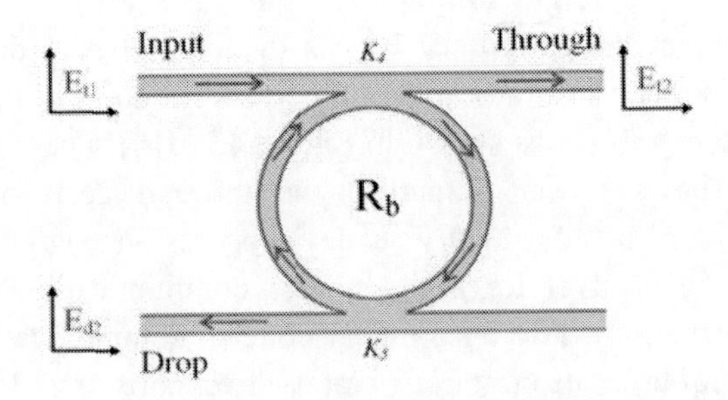

Figure 4.6. Schematic diagram of binary code recovery system.

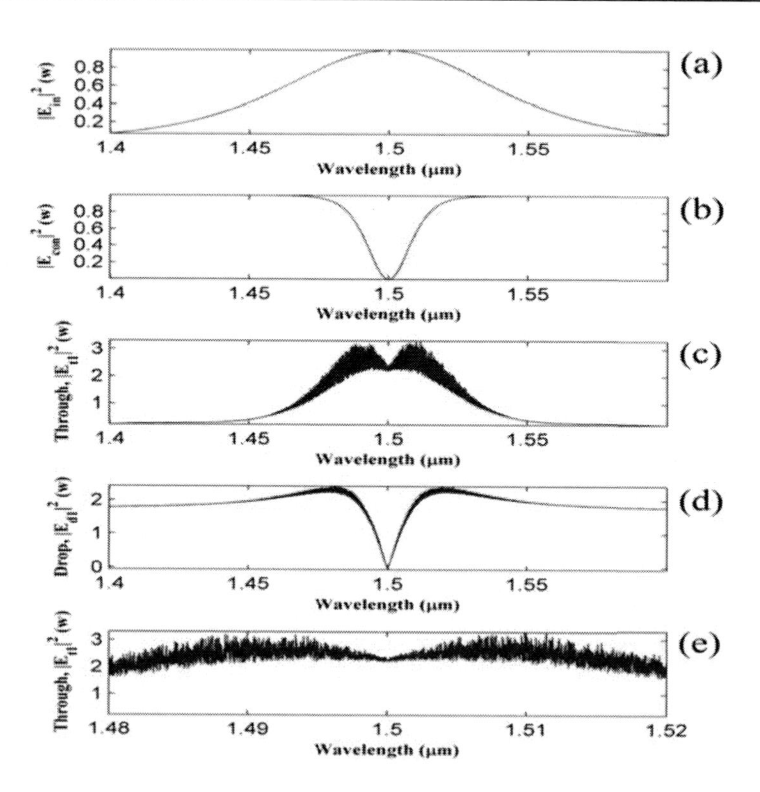

Figure 4.7. Simulation result of the light pulse generated at center wavelength λ_0 = *1.50 μm* for binary code suppression, where (a) $|E_{in}|^2$, (b) $|E_{con}|^2$, (c) $|E_{t1}|^2$, (d) $|E_{d1}|^2$ and (e) $|E_{t1}|^2$, where $R_r = 15\mu m$, $R_L = 15\mu m$, $R_{ad} = 200\mu m$ and $\alpha = 5 \times 10^{-5}\ dBmm^{-1}$.

Figure 4.7 shows the simulation result of the light pulse generated at center wavelength $\lambda_0 = 1.50\ \mu m$ for binary code suppression, where (a) $|E_{in}|^2$, (b) $|E_{con}|^2$, (c) $|E_{t1}|^2$, (d) $|E_{d1}|^2$ and (e) $|E_{t1}|^2$, where $R_r = 15\mu m$, $R_L = 15\mu m$, $R_{ad} = 200\mu m$ and $\alpha = 5 \times 10^{-5}\ dBmm^{-1}$. Here (a) is the input port for binary code suppression. By using bright soliton, the light pulse with 1W peak power is input into the input port. Figure 4.7(b) is the control port that uses a dark soliton light pulse with 1W peak power. The power output of the drop port for binary code suppression is as shown in Figure 4.7(d). Then, Figure 4.7(c) and Figure 4.7(e) shows power output of the through port for binary code suppression which is the binary code suppression signal that can be transmitted to the receiver for high security communication. The reference signal can be formed for reference binary code in communication. Moreover, the peak power output from through and drop ports are 2.3 and 3.2 W, respectively which is larger than the input light pulse.

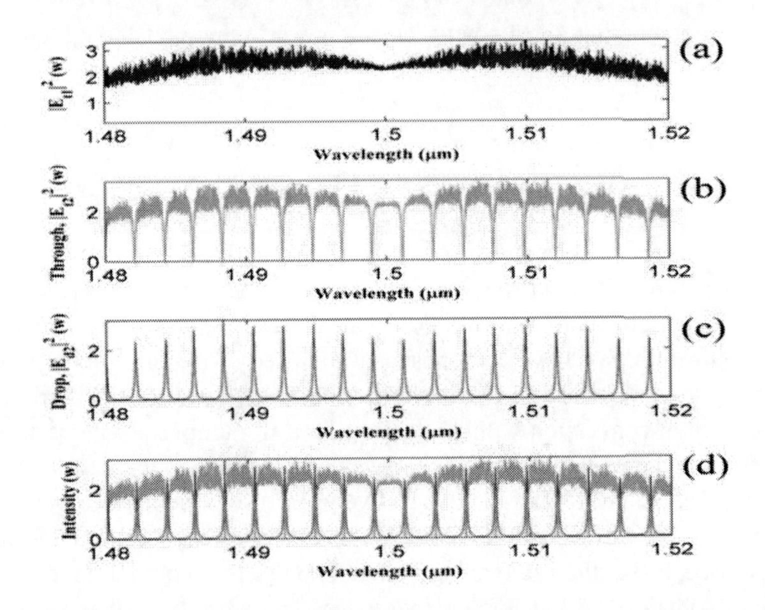

Figure 4.8. Simulation result of the light pulse generated at center wavelength $\lambda_0 = 1.50\ \mu m$ for binary code recovery, where (a) $|E_{t1}|^2$, (b) $|E_{t2}|^2$, (c) $|E_{d2}|^2$ and (d) compare $|E_{t2}|^2$ and $|E_{d2}|^2$, where $R_b = 50\mu m$ and $\alpha = 5 \times 10^{-5}\ dBmm^{-1}$.

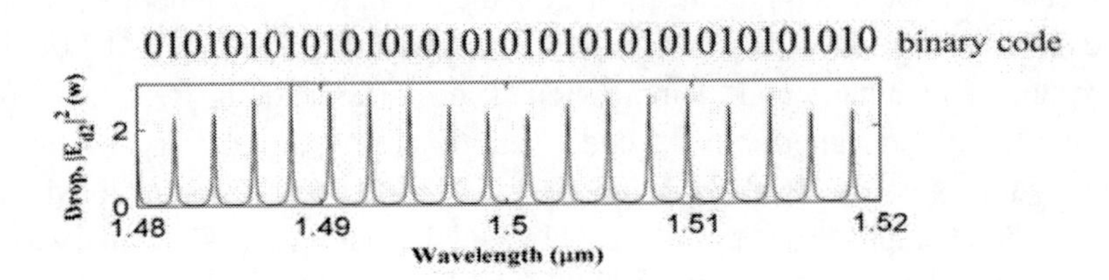

Figure 4.9. The binary code is obtained by the binary code recovery signals (Figure 4.8(c)).

Then, Figure 4.8 shows the simulation result of the light pulse generated at center wavelength $\lambda_0 = 1.50\ \mu m$ for binary code recovery, where (a) $|E_{t1}|^2$, (b) $|E_{t2}|^2$, (c) $|E_{d2}|^2$ and (d) compare $|E_{t2}|^2$ and $|E_{d2}|^2$, where $R_b = 50\mu m$ and $\alpha = 5 \times 10^{-5}\ dBmm^{-1}$. Here (a) is the input port for binary code recovery. The binary code suppression signal which looks like noisy signal is input into this port. It shows that this technique of communication is highly secured. Figure 4.8(b) and Figure 4.8(c) show the power output of the through port and drop port for binary code recovery, respectively. The power output from drop port represents the binary code, which is as shown in Figure 4.9. Figure 4.8(d) shows the comparing signals between the power outputs of though and drop ports, which have shown the signals relationship.

4.5. Conclusion

The authors have analyzed and described dark-bright soliton collision behavior within a *PANDA* ring resonator system. There is an interesting aspect called dark-bright soliton conversion that can be useful for random binary code application. We have also presented the used of the proposed system to generate the random codes, where the random binary codes using dark-bright optical soliton conversion within add/drop optical multiplexer can be formed. In order to achieve the required binary codes, first we set the dark and bight soliton pulses to be '0' and '1', respectively. By using the dark-bright soliton pulse trains, the dark and bright solitons can be separated by using PBS, which can be used to set the referencing codes. Finally, the required codes can be relatively obtained to the referencing codes. The advantage of the device is that the different binary code patterns can be randomly generated by changing the setting parameters, which can be available for security code applications

In this chapter, the authors also propose a new design for security application in optical communication by using the binary code suppression and recovery methods. By using the dark-bright soliton pair within a PANDA ring resonator. The binary code suppression is designed and the recovery code obtained. The designed binary code suppression and recovery system is complaint to possible creation. Moreover, the simulation results obtained have shown that the proposed system can indeed be achieved. The binary code suppression and recovery, in which the high security can be available for optical communication security. Future, it deserves to consider extending this work to real data suppression and recovery for realistic optical communication. Moreover, the development of the proposed technique for high security optical network can also be extended.

References

[1] Z. Jiang, D. E. Leaird and A. M. Weiner, "Line-by-line pulse shaping control for optical arbitrary waveform generation," *Opt. Express, vol.* 13, no. 25, pp. 10431-10439, Nov. 2005.

[2] B. Xia, L. R. Chen, P. Dumais, and C. L. Callender, "Ultrafast pulse train generation with binary code patterns using planar lightwave circuits," *Electron. Lett., vol.* 45, no. 19, pp. 1119-1120, Sep. 2006.

[3] A. Jain, I. A. Kostko, L. R. Chen, B. Xia, P. Dumais, and C. L. Callender, "Design and analysis of a 6-stage tunable MZI PLC for BPSK generation," in Proc. *50th IEEE Midwest Symposium on Circuits and Systems*, Montreal, QC, Canada (2007).

[4] P. Ou, Y. Zhang, and C.X. Zhang, "Optical generation of binary-phase-coded, direct-sequence ultra-wideband signals by polarization modulation and FBG-based multichannel frequency discriminator," *Opt. Express, vol.* 16, no. 7, pp. 5130-5135, Mar. 2008.

[5] T. Nishitani, T. Konishi, and K. Itoh, "Optical coding scheme using optical interconnection for high sampling rate and high resolution photonic analog-to-digital conversion," *Opt. Express, vol.* 15, no. 24, pp. 15812-15817, Nov. 2007.

[6] Y. Dai and J. Yao, "Optical generation of binary phase-coded direct-sequence UWB signals using a multichannel chirped fiber bragg grating," *J. Lightw. Technol., vol.* 26, no. 15, pp. 2513-2520, Aug. 2008.

[7] H. Huang, K. Xu, J. Li, J. Wu, X. Hong, and J. Lin, "UWB Pulse Generation and Distribution Using a NOLM Based Optical Switch," *J. Lightw. Technol., vol.* 26, no. 15, pp. 2635-2640, Aug. 2008.

[8] M. N. Akram, V. Kartashov, and Z. Tong, "Speckle reduction in line-scan laser projectors using binary phase codes," *Opt. Lett., vol.* 35, no. 3, pp. 444-446, Feb. 2010.

[9] K. Sarapat, N. Sangwara, K. Srinuanjan, P.P. Yupapin and N. Pornsuwancharoen, "Novel dark-bright optical solitons conversion system and power amplification," *Opt. Eng., vol.*48, pp. 045004, 2009.

[10] C. Teeka, P. Chaiyachet, P. P. Yupapin, N. Pornsuwancharoen, and T. Juthanggoon, "Soliton collision management in a microring resonator system," *Physics Procedia*, vol. 2, pp. 67-73, 2009.

[11] K. Uomwech, K. Sarapat and P.P. Yupapin, "Dynamic Modulated Gaussian Pulse Propagation within the Double PANDA Ring Resonator System," *Microwave and Optical Technology Letter*, vol. 52, no. 8, pp. 1818-1821, 2010.

[12] G. P. Agrawal, *Nonlinear fiber optics*, 4th edition, Academic Press, New York, 2007).

[13] F. M. Lee, C. L. Tsai, C. W. Hu, F. Y. Cheng, M. C. Wu, and C. C. Lin, "High-reliable and high-speed 1.3 μm complex-coupled distributed feedback buried-heterostructure laser diodes with Fe-doped

InGaAsP/InP hybrid grating layers grown by MOCVD," *IEEE Trans. on Elect. Devices, vol.* 55, no. 2, pp. 540-546, 2008.

[14] S. Tomofuji, S. Matsuo, T. Kakitsuka, and K. Kitayama, "Dynamic switching characteristics of InGaAsP/InP multimode interference optical waveguide switch," *Opt. Express, vol.* 17, no. 26, pp. 23380-23388, Dec. 2009.

[15] Y. Kokubun, Y. Hatakeyama, M. Ogata, S. Suzuki and N. Zaizen, "Fabrication technologies for vertically coupled microring resonator with multilevel crossing busline and ultracompact-ring radius," *IEEE J. Sel. Top. Quantum Electron.,* vol. 11, pp. 4-10, 2005.

[16] D. G. Rabus, M. Hamacher, U. Troppenz, and H. Heidrich, "Optical filters based on ring resonators with integrated semiconductor optical amplifiers in GaInAsP–InP," *IEEE J. Sel. Top. Quantum Electron*., vol. 8, no. 6, pp. 1405-1411, 2002.

[17] T. Phatharaworamet, C. Teeka, R. Jomtarak, S. Mitatha, and P. P. Yupapin, "Random Binary Code Generation Using Dark-Bright Soliton Conversion Control Within a PANDA Ring Resonator," *J. Lightw. Techn.,* vol. 28, no. 19, pp. 2804-2809, 2010.

Chapter 5

NANOELECTRONICS

5.1. INTRODUCTION

Optical device is become the interesting tool which can be involved in various applications, for instance, E/O (electrical/optical) and O/E(optical/electrical) signal converters, optical signal processing, optical sensor, optical communication and medicine, etc. More interesting applications have been appeared, especially, when the device dimension is reached the micro/nano scale regime, in which many aspects of investigation become the challenge. In this paper, we propose the new design of the electronic arithmetic and logic circuit for ultrahigh speed information processing, in which the optical signal in electrical domain can be formed by using the optical converter and anticipated to confront the speed and bandwidth limitation. In operation, a binary half adder/substractor is the most importance operation of two digits because the half adder/substractor can be used to implement a full adder/substractor. However, many optical arithmetic and logic have been proposed, such as semiconductor optical amplifier (SOA) [1-3], quantum dot [4, 5], TOAD based interferometer device [6], cascaded microring resonators [7], all-optical arithmetic unit [8, 9]. However, the searching of new techniques remains, therefore in this paper, we propose the simultaneous operation of half adder/subtractor arithmetic and logic gate based on dark-bright soliton conversion behavior, in which the coincidence dark and bright soliton can be separated after propagating into coupler at $\pi/2$ phase shift device (an optical coupler). The proposed scheme is based on a 1 bit binary compared to the complex logic circuits, which can be compared to any 2 bits, when logic '0' and '1' using the dark and bright soliton, respectively.

5.2. Dark-Bright Soliton Conversion

In the operation, the dark-bright soliton conversion using a ring resonator optical channel dropping filter (OCDF)[10, 11] is composed of two set of coupled waveguide, as shown in Figure 5.1(a) and 5.1(b), when for convenience, Figure 5.1(b) is replaced by Figure 5.1(a). The relative phase of the two output light signals after coupling into the optical coupler is π/2 before coupling into the ring and the input bus, respectively. This mean that the signals coupled into the drop and through ports are acquired a phase of π with respect to the input port signal. In application, if we performed the coupling coefficients appropriately, the field coupled into the through port on resonance would completely extinguish the resonant wavelength, and all power would be coupled into the drop port. When the dark-bright conversion is show in Eqs. (5.1) - (5.8).

$$E_{ra} = -j\kappa_1 E_i + \tau_1 E_{rd}, \tag{5.1}$$

$$E_{rb} = \exp(j\omega T/2)\exp(-\alpha L/4)E_{ra}, \tag{5.2}$$

$$E_{rc} = \tau_2 E_{rb} - j\kappa_2 E_a, \tag{5.3}$$

$$E_{rd} = \exp(j\omega T/2)\exp(-\alpha L/4)E_{rc}, \tag{5.4}$$

$$E_t = \tau_1 E_i - j\kappa_1 E_{rd}, \tag{5.5}$$

$$E_d = \tau_2 E_a - j\kappa_2 E_{rb}, \tag{5.6}$$

where E_i is the input field, E_a is the added(control) field, E_t is the throughput field, E_d is the dropped field, $E_{ra}...E_{rd}$ are the fields in the ring at the point $a...d$, κ_1 is the field coupling coefficient between the input and the ring, κ_2 is the field coupling coefficient between the ring and the output bus, L is the circumference of the ring ($2\pi R$), T is the time taken for one round trip, $T = Ln_{eff}/c$, and α is the power loss in the ring per unit length. We assume that lossless coupling, i.e. $\tau_{1,2} = \sqrt{1-\kappa_{1,2}^2}$. The output power/intensities at the drop port and through port are given by

$$|E_d|^2 = \left| \frac{-\kappa_1\kappa_2 A_{1/2}\Phi_{1/2}}{1-\tau_1\tau_2 A\Phi} E_i + \frac{\tau_2 - \tau_1 A\Phi}{1-\tau_1\tau_2 A\Phi} E_a \right|^2 . \quad (5.7)$$

$$|E_t|^2 = \left| \frac{\tau_2 - \tau_1 A\Phi}{1-\tau_1\tau_2 A\Phi} E_i + \frac{-\kappa_1\kappa_2 A_{1/2}\Phi_{1/2}}{1-\tau_1\tau_2 A\Phi} E_a \right|^2 . \quad (5.8)$$

where $A_{1/2} = \exp(-\alpha L/4)$(the half-round-trip amplitude), $A = A_{1/2}^2$, $\Phi_{1/2} = \exp(j\omega T/2)$ (the half-round-trip phase contribution), and $\Phi = \Phi_{1/2}^2$.The input and control fields at the input and add ports are formed by the dark-bright optical soliton [12, 13] as shown in Eqs. (5.9) – (5.10).

$$E_{in}(t) = A_0 \sec h\left[\frac{T}{T_0}\right] \exp\left[\left(\frac{z}{2L_D}\right) - i\omega_0 t\right] \quad (5.9)$$

$$E_{in}(t) = A_0 \tanh\left[\frac{T}{T_0}\right] \exp\left[\left(\frac{z}{2L_D}\right) - i\omega_0 t\right] \quad (5.10)$$

where A and z are optical field amplitude and propagation distance, respectively. T is soliton pulse propagation time in a frame moving at the group velocity $T = t - \beta_1 z$,where β_1 and β_2 are the coefficients of the linear and second-order terms of Taylor expansion of the propagation constant. $L_D = T_0^2 / |\beta_2|$ is the dispersion length of the soliton pulse. T_0 in the equation is the initial soliton pulse width, where t is the soliton phase shift time, and the frequency shift of the soliton is ω_0 . This solution describes a pulse that keeps its temporal width invariance as it propagates, and thus is called a temporal soliton. When a soliton peak intensity $(\beta / \Gamma T_0^2)$ is given, then T_0 is known. For the soliton pulse in the nanoring device, a balance should be achieved between the dispersion length (L_D) and nonlinear length $L_{NL} = (1/\Gamma\phi_{NL})$, where $\Gamma = n_2 k_0$, is the length scale over which dispersive or nonlinear effects make the beam become wider or narrower. For a soliton pulse, there is a balance between dispersion and nonlinear lengths, hence $L_D = L_{NL}$.

5.3. SIMULTANEOUS ARITHMETIC OPERATION

In operation, the binary arithmetic is performed by the same decimal arithmetic way, which is presented by the logic gate operation. In the design, for simplicity, the multiple input ports are required to form the device functions, where first of all, the required half adder/subtractor truth table is given and shown in Table 5.1. For the half adder/subtractor with two binary inputs, the simplified Boolean equation is obtained, in which the sum of product for each output is given by Eqs. (5.11) - (5.14) for half adder and half subtractor, respectively. The simplified output of Sum and Difference can be also implemented with XOR gate as shown in Figure 5.2, in which the addition and subtraction operations can be combined into one circuit with one common binary adder.

$$Sum = \overline{X}Y + X\overline{Y} \tag{5.11}$$

$$Carry = XY \tag{5.12}$$

$$Difference = \overline{X}Y + X\overline{Y} \tag{5.13}$$

$$Borrow = \overline{X}Y \tag{5.14}$$

An all-optical half adder/subtractor system is as shown in Figure 5.1(c). When the input and control light pulse trains are input into the first add/drop optical filter (MRR1), in which the dark soliton (logic ‘0’) or the bright soliton (logic ‘1’) is formed the within the device. Firstly, the dark soliton is converted to be dark and bright soliton via the add/drop optical filter [14], which they can be seen at the through and drop ports with π phase shift [15], and then it can form inverter gate (NOT gate), respectively. By using the add/drop optical filter (MRR2 and MRR3), both input signal are generated by the first stage add/drop optical filter. In the next procedure, the input data “Y” with logic “0” (dark soliton) and logic “1” (bright soliton) are added into both add ports, the dark-bright soliton conversion, in which the π phase shift is operated again. For large scale (Figure 5.1(c)), results obtained are simultaneously seen by $D_2, D_3, T_2,$ and T_3 at the drop and through ports for optical logic operation.

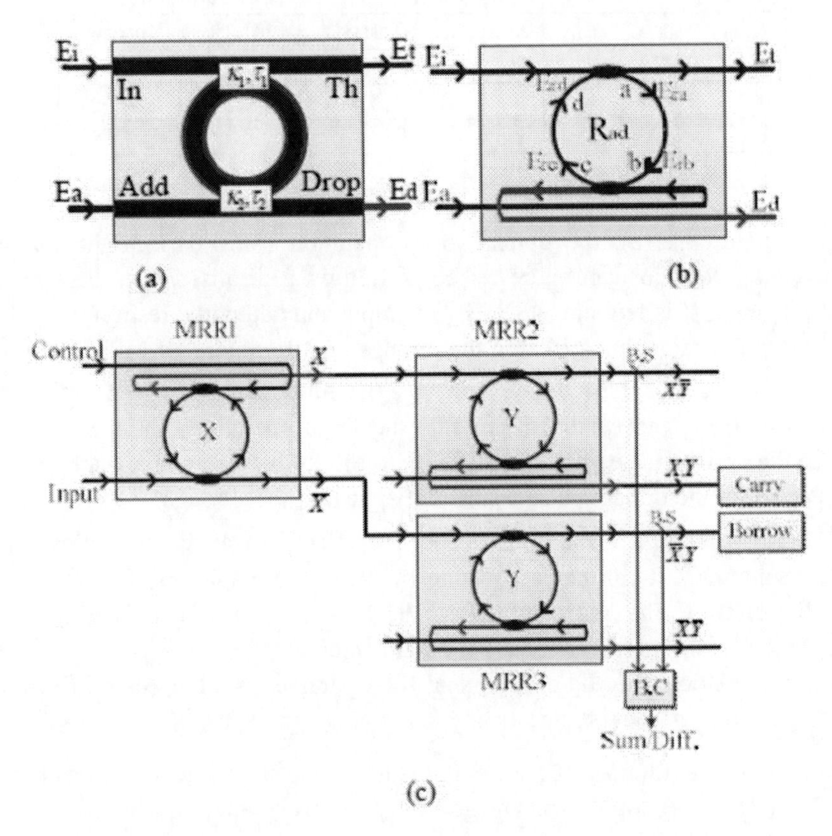

Figure 5.1. A schematic diagram of a simultaneous optical logic device.

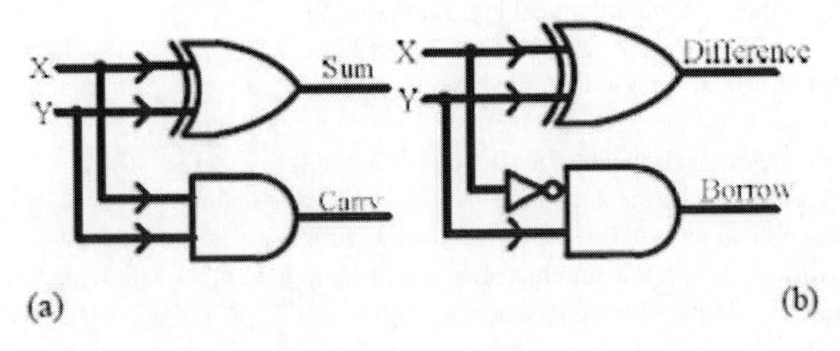

Figure 5.2. (a) a half adder, (b) a half subtractor.

Table 5.1. Truth table of the half adder/subtractor

Input		Half Adder		Half Subtractor	
X	Y	Sum	Carry	Diff.	Borrow
0	0	0	0	0	0
0	1	1	0	1	1
1	0	1	0	1	0
1	1	0	1	0	0

In Figure 5.1(c), the optical logic operation using dark-bright soliton conversion behavior can be described as following details. When the optical pulse train X, Y is fed into MRR2 by the input and add ports, respectively, the optical pulse trains that appear at the through and drop ports of MRR2 will be $X.\overline{Y}$ and $X.Y$, respectively, whereas the aforementioned assumption is provided. Here, the symbol represents the logic operation AND. Similarly, when the optical pulse train X, Y is fed into MRR3 by the input and add ports, respectively, the optical pulse trains that appear at the through and drop ports of MRR3 will be $\overline{X}.Y$ and $\overline{X}.\overline{Y}$, respectively. To generate the all-optical half adder/subtractor, it can be easily done by using beam splitters (BS) or beam combiner (BC). The beam splitters to be used here are not polarizing, and reflect (and transmit) 50% of the light that is incident.

In simulation, the add/drop optical filter parameters are used and fixed to be $\kappa_s = 0.5, R_{ad} = 1.5\mu m$[16], $A_{eff} = 0.25\mu m^2, \alpha = 0.05 dBmm^{-1}$ for all add/drop optical filters in the system. Results of the simultaneous optical logic XOR and XNOR logic gate are generated by using the dark-bright soliton conversion with wavelength center at $\lambda_0 = 1.50\mu m$, pulse width $35 fs$ and the input data logic "0" (dark soliton) and logic "1" (bright soliton). In Figure 5.3, the simultaneous output optical logic gate is seen.

Case 1: When the simultaneous output logic gate input data logic "00" is added, the obtained output optical logic is "0001" [see Figure 5.3 (a)].

Case 2: When the simultaneous output logic gate input data logic "01" is added, the output optical logic "0010" is formed [see Figure 5.3(b)].

Case 3: When the simultaneous output logic gate input is "10" added, the output optical logic "1000" if formed [see Figure 5.3(c)].

Case 4: When the simultaneous output logic gate input data logic "11" is added, we found the output optical logic "0100" is obtained [see Figure 5.3(d)].

The simultaneous all-optical half adder/substractor and logic is concluded in Table 5.2. We found that the output logic in the drop port, D_2, D_3 are optical logic XNOR gate, whereas the output logic in the through ports, T_2 and T_3 are optical logic XOR gates, where the simultaneously logic gate operations can be formed for half adder/subtractor arithmetic, respectively .

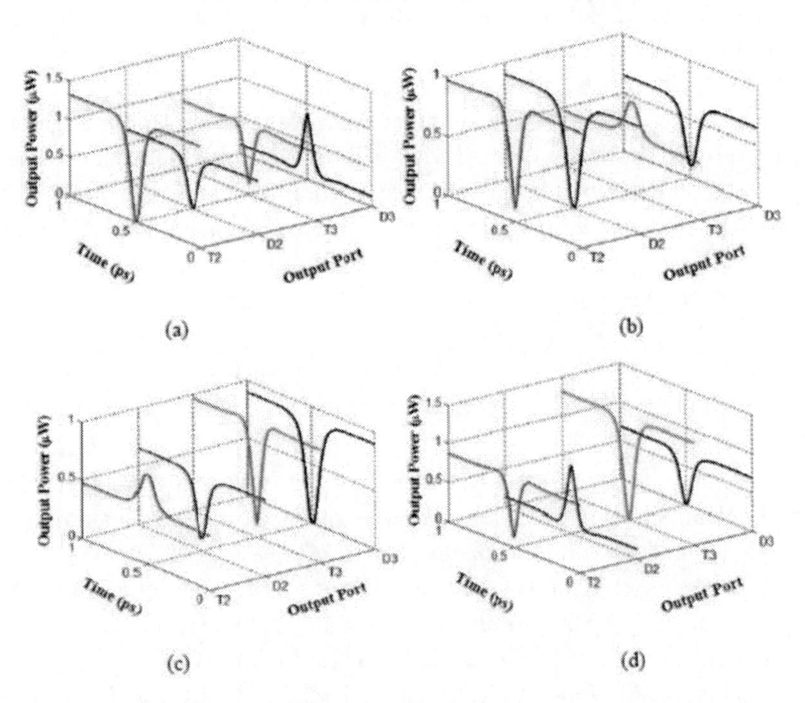

Figure 5.3. Simulation results of the output logic gates when the input logic states are (a) 'DD', (b) 'DB', (c) 'BD' and (d) 'BB', respectively.

Table 5.2. Conclusion output of the optical logic gate

X	Y	(T_2) $X.\overline{Y}$	(D_2) $X.Y$	(T_3) $\overline{X}.Y$	(D_3) $\overline{X}.\overline{Y}$	Sum/Diff (XOR) $X.\overline{Y}+\overline{X}.Y$	XNOR $X.Y+\overline{X}.\overline{Y}$
D	D	D	D	D	B	D	B
D	B	D	D	B	D	B	D
B	D	B	D	D	D	B	D
B	B	D	B	D	D	D	B

5.4. Conclusion

We have shown that the simultaneous logic gate operation between half adder and subtractor can be formed by using the dark-bright soliton conversion system via the add/drop optical filter, which is formed by all-optical circuit using microring and nanoring device structures, in which the advantage of the simultaneous logic gate operation between the half adder and subtractor arithmetic are seen. By using the dark-bright soliton conversion system, the input data logic '0' (dark soliton) and logic '1' (bright soliton) is established, in which the logic status results simultaneously at the drop and through ports, respectively. In application, this device will be the great component to use for digital photonic circuit design and recognized as the simple and flexible system for logic switching system, which can be extended and implemented for any higher number of input digits by a proper incorporation of dark-bright soliton conversion control base optical switches.

References

[1] J. Cong, X. Zhang and D. Huang, A propose for two-input arbitrary Boolean logic gates using single semiconductor optical amplifier by picosecond pulse injection, *Opt. Exp.*, 17(2009)7725-7730.

[2] S.H. Kim, J.H. Kim, B.G. Yu, Y.T. Byun, Y.M. Jeon, S.Lee, D.H. Woo, All-optical NAND gate using cross-gain modulation in semiconductor optical amplifiers, *Electronic Lett.*, 41(2005)1027-1028.

[3] J. Hun, Y. M. John, Y. T. Byun, S. Lee, D. H. Woo, S. H. Kim, All-Optical XOR Gate Using Semiconductor Optical Amplifiers Without Additional Input Beam, Photonics Technology *Lett., IEEE,* 14(2002)1436-1438.

[4] S. Ma, Z. Chen, H. Sun, and K. Dutta, High speed all optical logic gates based on quantum dot semiconductor optical amplifiers, *Opt. Exp.,* 18(2010)6417-6422.

[5] T. Kawazoe, K. Kobayashi, K. Akahane, M. Naruse, N. Yamamoto, and M. Ohtsu, Demonstration of nanophotonic NOT gate using near-field optically coupled quantum dot, *Appl. Phys. B.*, 84(2006)243-246.

[6] J. N. Roy, and D. K. Gayen, Integrated all-optical logic and arithemetic operation with the help of a TOAD-based interferometer device-alternative approach, *Appl. Opt.*, 46(2007)5304-5310.

[7] L. Zhang, R. Ji, L. Jia, L. Yang, P. Zhou, Y. Tiam, P. Chen, Y. Lu, Demonstration of directed XOR/XNOR logic gates using two cascaded microring resonators, *Opt. Lett.*, 35(10)(2010) 1620-1622.

[8] D. K. Gayen , J. N. Roy, All-optical arithmetic unit with the help of terahertz-optical-asymmetric-demultiplexer-based tree architecture, *Appl. Opt.,* 47(2008)933-943.

[9] J. N. Roya., A. K. Maitib, D. Samantac, S. Mukhopadhyayc, Tree-net architecture for integrated all-optical arithmetic operations and data comparison scheme with optical nonlinear material, *Optical Switching and Networking.,* 4(2007)231-237.

[10] P. P. Absil, J. V. Hryniewicz, B. E. Little, F. G. Johnson, and P.-T. Ho, Vertically coupled microring resonators using polymer wafer bonding, *IEEE Photon. Technol. Lett.*, 13(2001)49-51.

[11] R. Grover, P. P. Absil, V. Van, J. V. Hryniewicz, B. E. Little, O. S. King, L. C. Calhoun, F. G. Johnson, and P.-T. Ho, Vertically coupled GaInAsP-InP microring resonators, *Opt. Lett.*, 26(2001)506-508.

[12] S. Mitatha, N. Pornsuwancharoen, P. P. Yupapin, A simultaneous short-wave and millimeter-wave generation using a soliton pulse with in a nano-waveguide, *IEEE Photon. Technol. Lett.*,13(2009)932-934.

[13] V. Van, T. A. Ibrahim, P. P. Absil, F. G. Johnson, R. Grover, Optical signal processing using nonlinear semiconductor micro ring resonators, *IEEE J. of Sel. Top. in Quantum Electron.*, 8(2002)705-713.

[14] S. Mitatha, N. Chaiyasoonthorn, P. P. Yupapin, Dark-bright optical solitons conversion via an optical add/drop filter, *Microw. and Opt. Technol. Lett.,* 51(2009) 2104-2107.

[15] S. Mookherjea and M. A. Schneider, The nonlinear microring add-drop filter, *Opt. Exp.*, 16(2008)15130-15136.

[16] Q. Xu, D. Fattal, R. G. Beausoleil, Silicon microring resonators with 1.5-μm radius, *Opt. Exp.*, 16(2008)4309-4315.

Chapter 6

HYBRID TRANSCEIVER

6.1. INTRODUCTION

Nano communication and networking has become the challenge and interesting aspect of research and investigating recently, where mostly, the microscale and nanoscale devices are the basic components for such systems. Microring/nanoring resonator has also become an interesting device, which has been widely studied and investigated in many research areas [1-4], which one of them has shown that the use of ring resonator device can form the interesting aspect, for instance, the trapped/retrieved photon/atom in/from a microring device system may be used to bring photons/atoms for long distance transmission. Several researchers have shown that the dynamic optical trapping tools can be formed by controlling the dark-bright soliton behaviors within a semiconductor add/drop multiplexer (ring resonator), which have been clearly investigated [5-8]. In those investigations, the high optical field is configured as the optical tweezers or potential wells [9, 10], which is available for photon/atom trapping and transportation. However, the searching of new suitable technique for photons/atoms trapping and transportation remains, in which there are plenty of rooms required to investigate and accommodate.

Optical tweezers technique has become a powerful tool for manipulation of micrometer-sized particles in three spatial dimensions. Initially, the useful static tweezers are recognized, and the dynamic tweezers are now realized in practical works [11-13]. Recently, Schulz et al [14] have shown that the transfer of trapped atoms between two optical potentials could be performed. In principle, an optical tweezers use forces exerted by intensity gradients in the strongly focused beams of light to trap and move the microscopic volumes of matters. Moreover, the other combination of force is induced by the interaction between photons, which is caused by the photon scattering effects. In application, the field intensity can be adjusted and tuned to form the suitable trapping potential, in which the desired gradient field and scattering force can be formed the suitable trapping force. Hence, the appropriated force can be configured for the transmitter/receiver part, which can be performed the long distance transportation.

In this chapter, the dynamic optical vortices are generated using a dark soliton, bright soliton and Gaussian pulse propagating within an add/drop optical multiplexer incorporating two nanoring resonators (PANDA ring resonator). The dynamic behaviors of solitons and Gaussian pulses are analyzed and described. To increase the channel multiplexing, the dark solitons with slightly different wavelengths are controlled and amplified within the tiny system. The trapping force stability is simulated and seen when the Gaussian pulse is used to control via the add(control) port. In application, the optical vortices (dynamic tweezers) can be used to store (trap) photon, atom, molecule, DNA, ion, or particle, which can perform the dynamic tweezers. By using the hybrid devices, the hybrid transceiver and repeater can be integrated and performed the required functioned by using a single system, here, the use of the transceiver and repeater to form the hybrid communication of those microscopic volumes of matters in the nanoscale regime can be realized.

6.2. THEORY

In operation, the optical tweezers use forces that are exerted by the intensity gradients in the strongly focused beams of light to trap and move the microscopic volumes of matters, in which the optical forces are customarily defined by the relationship [15].

$$F = \frac{Qn_m P}{c} \tag{6.1}$$

where Q is a dimensionless efficiency, n_m is the index of refraction of the suspending medium, c is the speed of light, and P is the incident laser power, measured at the specimen. Q represents the fraction of power utilized to exert force. For plane waves incident on a perfectly absorbing particle, Q is equal to

1. To achieve stable trapping, the radiation pressure must create a stable, three-dimensional equilibrium. Because biological specimens are usually contained in aqueous medium, the dependence of F on n_m can rarely be exploited to achieve higher trapping forces. Increasing the laser power is possible, but only over a limited range due to the possibility of optical damage. Q itself is therefore the main determinant of trapping force. It depends upon the NA(numerical aperture), laser wavelength, light polarization state, laser mode structure, relative index of refraction, and geometry of the particle.

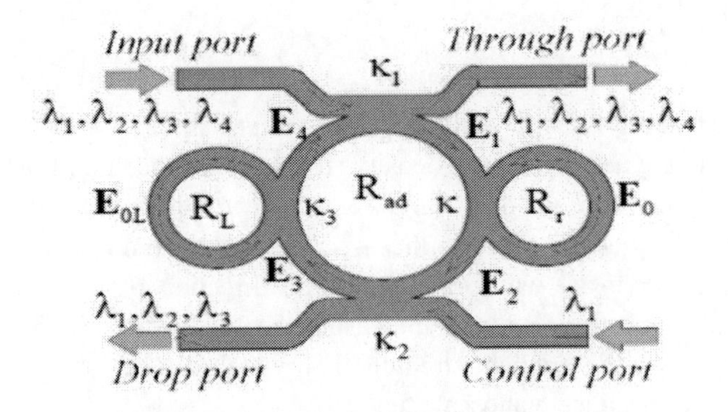

Figure 6.1. Schematic diagram of a proposed hybrid transceiver.

Furthermore, in the Rayleigh regime, trapping forces decompose naturally into two components. Since, in this limit, thc clcctromagnetic field is uniform across the dielectric, particles can be treated as induced point dipoles. The scattering force is given by

$$F_{scatt} = n_m \frac{\langle S \rangle \sigma}{c}, \tag{6.2}$$

where

$$\sigma = \frac{8}{3}\pi (kr)^4 r^2 \left(\frac{m^2 - 1}{m^2 + 2}\right)^2 \tag{6.3}$$

is the scattering cross section of a Rayleigh sphere with radius r. S is the time-averaged Poynting vector, n is the index of refraction of the particle, $m = n / n_m$ is the relative index, and $k = 2\pi n_m / \lambda$ is the wave number of the light. Scattering force is proportional to the energy flux and points along the direction of propagation of the incident light. The gradient force is the Lorentz force acting on the dipole induced by the light field. It is given by

$$F_{grad} = \frac{\alpha}{2}\nabla\left\langle E^2\right\rangle, \tag{6.4}$$

where

$$\alpha = n_m^2 r^3\left(\frac{m^2-1}{m^2+2}\right) \tag{6.5}$$

is the polarizability of the particle. The gradient force is proportional and parallel to the gradient in energy density (for $rn > 1$). The large gradient force is formed by the large depth of the laser beam, in whcih the stable trapping requires that the gradient force in the $-\hat{z}$ direction, where it is a dark soliton valley in this proposal, which is against the direction of incident light, be greater than the scattering force. Increasing the NA decreases the focal spot size and increases the gradient strength [16], which can be formed within the tiny system, for instance, nanoscale device.

We are looking for the system that can generate the dynamic tweezers (optical vortices), in which the microscopic volume can be trapped and transmission via the communication link. Firstly, the stationary and strong pulse that can propagate within the dielectric material (waveguide) for period of time is required. Moreover, the gradient field is an important property required in this case. Therefore, a dark soliton is satisfied and recommended to perform those requirements. Secondly, we are looking for the device that optical tweezers can propagate and form the long distance link, in which the gradient field (force) can be transmitted and received by using the same device. Here, the add/drop multiplexer in the form of a PANDA ring resonator which is well known and is introduced for this proposal, as shown in Figs. 1 and 2. To form the multi function operations, for instance, control, tune, amplify, the additional pulses are bright soliton and Gaussian pulse introduced into the system. The input optical field (E_{in}) and the add port optical field (E_{add}) of the dark soliton, bright soliton and Gaussian pulses are given by [17], respectively.

$$E_{in}(t) = A_0 \tanh\left[\frac{T}{T_0}\right]\exp\left[\left(\frac{z}{2L_D}\right) - i\omega_0 t\right] \tag{6.6a}$$

$$E_{control}(t) = A\,\mathrm{sech}\left[\frac{T}{T_o}\right]\exp\left[\left(\frac{Z}{2L_D}\right) - i\omega_o t\right] \tag{6.6b}$$

$$E_{control}(t) = E_0 \exp\left[\left(\frac{z}{2L_D}\right) - i\omega_0 t\right] \tag{6.6c}$$

where A and z are the optical field amplitude and propagation distance, respectively. T is a soliton pulse propagation time in a frame moving at the group velocity, $T=t-\beta_1 z$, where β_1 and β_2 are the coefficients of the linear and second-order terms of Taylor expansion of the propagation constant. $L_D = T_0^2/|\beta_2|$ is the dispersion length of the soliton pulse. T_0 in equation is a soliton pulse propagation time at initial input (or soliton pulse width), where t is the soliton phase shift time, and the frequency shift of the soliton is ω_0. This solution describes a pulse that keeps its temporal width invariance as it propagates, and thus is called a temporal soliton. When a soliton of peak intensity $(|\beta_2/\Gamma T_0^2|)$ is given, then T_0 is known. For the soliton pulse in the microring device, a balance should be achieved between the dispersion length (L_D) and the nonlinear length ($L_{NL}=1/\Gamma\phi_{NL}$) . Here $\Gamma=n_2k_0$, is the length scale over which dispersive or nonlinear effects makes the beam become wider or narrower. For a soliton pulse, there is a balance between dispersion and nonlinear lengths. Hence $L_D = L_{NL}$. For Gaussian pulse in Eq. (6.6c), E_0 is the amplitude of optical field.

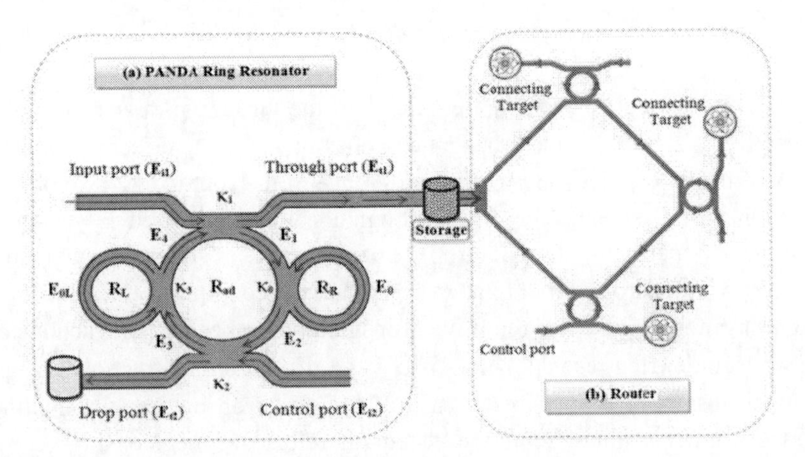

Figure 6.2. Schematic diagram of a hybrid transceiver and a router.

When light propagates within the nonlinear medium, the refractive index (n) of light within the medium is given by

$$n = n_0 + n_2 I = n_0 + \frac{n_2}{A_{eff}} P, \tag{6.7}$$

with n_0 and n_2 as the linear and nonlinear refractive indexes, respectively. I and P are the optical intensity and the power, respectively. The effective mode core area of the device is given by A_{eff}. For the add/drop optical filter design, the effective mode core areas range from 0.50 to 0.10 μm^2, in which the parameters were obtained by using the related practical material parameters (InGaAsP/InP)[18]. When a dark soliton pulse is input and propagated within a add/drop optical filter as shown in Figure 6.1, the resonant output is formed.

In order to retrieve the required signals, we propose to use the add/drop device with the appropriate parameters. This is given in the following details. The optical circuits of ring resonator add/drop filters for the through port and drop port can be given by Eqs. (6.8) and (6.9), respectively[20].

$$\left|\frac{E_t}{E_{in}}\right|^2 = \frac{\left[(1-\kappa_1)+(1-\kappa_2)e^{-\alpha L} - 2\sqrt{1-\kappa_1}\cdot\sqrt{1-\kappa_2}e^{-\frac{\alpha}{2}L}\cos(k_n L)\right]}{\left[1+(1-\kappa_1)(1-\kappa_2)e^{-\alpha L} - 2\sqrt{1-\kappa_1}\cdot\sqrt{1-\kappa_2}e^{-\frac{\alpha}{2}L}\cos(k_n L)\right]} \tag{6.8}$$

$$\left|\frac{E_d}{E_{in}}\right|^2 = \frac{\kappa_1\kappa_2 e^{-\frac{\alpha}{2}L}}{1+(1-\kappa_1)(1-\kappa_2)e^{-\alpha L} - 2\sqrt{1-\kappa_1}\cdot\sqrt{1-\kappa_2}e^{-\frac{\alpha}{2}L}\cos(k_n L)} \tag{6.9}$$

Here E_t and E_d represent the optical fields of the through port and drop ports, respectively. $\beta = kn_{\text{eff}}$ is the propagation constant, n_{eff} is the effective refractive index of the waveguide, and the circumference of the ring is $L=2\pi R$, with R as the radius of the ring. The filtering signal can be managed by using the specific parameters of the add/drop device, and the required signals can be retrieved via the drop port output. κ_1 and κ_2 are the coupling coefficients of the add/drop filters, $k_n=2\pi/\lambda$ is the wave propagation number for in a vacuum, and the waveguide (ring resonator) loss is $\alpha = 0.5$ dBmm^{-1}. The fractional coupler intensity loss is $\gamma = 0.1$. In the case of the add/drop device, the nonlinear refractive index is not effect to the system, therefore, it is neglected.

From Eq. (6.8), the output field (E_{t1}) at the through port is given by

$$E_{t2} = AE_{i1} - BE_{i2}e^{-\frac{\alpha L}{2} - jk_n\frac{L}{2}} - \left[\frac{CE_{i1}e^{-\frac{\alpha L}{2} - jk_n\frac{L}{2}} + DE_{i2}e^{-\frac{3\alpha L}{2} - jk_n\frac{3L}{2}}}{1 - YE_0E_{0L}e^{-\alpha L - jk_nL}}\right] \tag{6.10}$$

where $A = \sqrt{(1-\gamma_1)(1-\gamma_2)}$, $B = \sqrt{(1-\gamma_1)(1-\gamma_2)\kappa_1(1-\kappa_2)}E_{0L}$, $C = \kappa_1(1-\gamma_1)\sqrt{(1-\gamma_2)\kappa_2}E_0E_{0L}$, $D = (1-\gamma_1)(1-\gamma_2)\sqrt{\kappa_1(1-\kappa_1)\kappa_2(1-\kappa_2)}E_0E_{0L}^2$ and $Y = \sqrt{(1-\gamma_1)(1-\gamma_2)(1-\kappa_1)(1-\kappa_2)}$

The electric fields E_0 and E_{0L} are the field circulated within the nanoring at the right and left side of add/drop optical filter.

The power output (P_{t1}) at through port is written as

$$P_{t1} = |E_{t1}|^2. \tag{6.11}$$

The output field (E_{t2}) at drop port is expressed as

$$E_{t2} = \sqrt{(1-\gamma_2)(1-\kappa_2)}E_{i2} - \left[\frac{\sqrt{(1-\gamma_1)(1-\gamma_2)\kappa_1\kappa_2}E_0E_{i1}e^{-\frac{\alpha L}{2} - jk_n\frac{L}{2}} + XE_0E_{0L}E_{i2}e^{-\alpha L - jk_nL}}{1 - YE_0E_{0L}e^{-\alpha L - jk_nL}}\right] \tag{6.12}$$

where $X = (1-\gamma_2)\sqrt{(1-\gamma_1)(1-\kappa_1)\kappa_2(1-\kappa_2)}$, $Y = \sqrt{(1-\gamma_1)(1-\gamma_2)(1-\kappa_1)(1-\kappa_2)}$

The power output (P_{t2}) at drop port is

$$P_{t2} = |E_{t2}|^2. \tag{6.13}$$

6.3. Hybrid Transceiver and Repeater

Simulation results of trapped multi photons within the optical vortices are as shown in Figure 6.3, the coupling coefficients are given as $\kappa_0 = 0.1$, $\kappa_1 = 0.35$, $\kappa_2 = 0.1$ and $\kappa_3 = 0.2$, respectively. The ring radii are $R_{add} = 10\mu m$, $R_{right} = 1.5\ \mu m$ and $R_{left} = 1.5\ \mu m$, in which the evidence of the practical device was reported by the authors in reference [21, 22]. A_{eff} are 0.50, 0.25 and 0.25 μm^2. From Figure 6.3(a)-(e), shows the results of four potential wells (tweezers) with four different center wavelengths $\lambda_1 = 1.4\mu m$, $\lambda_2 = 1.45\mu m$, $\lambda_3 = 1.5\mu m$ and $\lambda_4 = 1.6\mu m$ for multi photons trapping using dynamic potential wells, in which they are controlled by the Gaussian signal with 1 W peak power at the control port, where the signals are (a) $|E_1|^2$, (b) $|E_2|^2$, (c) $|E_3|^2$, (d) $|E_4|^2$, (e) through port and (f) drop port signals, the dark soliton and Gaussian pulse are used as the input and control ports respectively. The important aspect is that the tunable receiver can be operated by tuning (controlling) the add (control) port input signal, in which the required number of microscopic volume (atom/photon/molecule) can be obtained via the drop port, otherwise, they are propagated within a PANDA device for a period of time before collapsing in the device. However, the use of amplification concept can keep the system being active in the same way as a hybrid memory, whereas the microscopic volume can be stored.

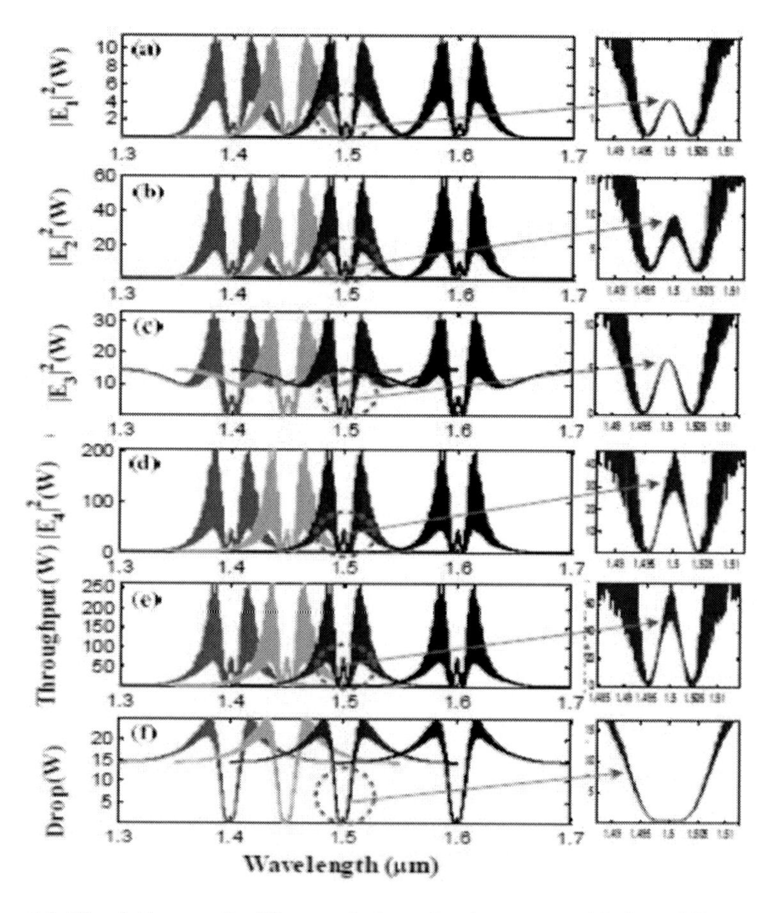

Figure 6.3. Simulation result of four optical vortices/tweezers with four different center wavelengths.

More results of are as shown in Figure 6.4, where in this figure the multi tweezers can be generated with slightly different wavelengths, and the tunable gradient fields are obtained by varying the coupling coefficients, which can be formed in the fabrication process. The trapped multi photons(a) and tweezers(b) at the drop port are seen, in which the tunable coupling coefficients(κ_2) are (1) 0.1, (2) 0.35, (3) 0.6 and (4) 0.75, respectively, the other coupling coefficients are given as $\kappa_0 = 0.5$, $\kappa_1 = 0.35$, and $\kappa_3 = 0.1$. The multi photons are trapped with center wavelengths $\lambda_1 = 1.4\mu m$, $\lambda_2 = 1.45\mu m$, $\lambda_3 = 1.5\mu m$ and $\lambda_4 = 1.6\mu m$, where (a) $|E_1|^2$, (b) $|E_2|^2$, (c) $|E_3|^2$, (d) $|E_4|^2$ and (e) are outputs from the through port, and (f) are the output at the drop port, respectively. We found that the multi photons can be trapped and seen by controlling the second coupler parameter (coupling coefficient) of the add/drop device, moreover, the trapped multi photons can be amplified and tuned as shown in Figure 6.4(a) – (f), in which the multi trapped photons with different coupling coefficients (κ_2) are increased and seen, whereas the trapping depth is increased. More simulation results of the dynamic optical vortices within the PANDA device are as shown in Figure 6.5. In this case the bright soliton is input into the control port, and the received atoms/molecules are seen as shown in Figure 5(a)-5(d).

6.3.1. Hybrid Transceiver

In application, the transmitter and receiver are formed by using through port and drop port outputs respectively, whereas such the device can be operated both functions by the single device, which is called a hybrid transceiver.

The required amount of microscopic volume can be controlled by using the hybrid transceiver and forming the transmission link via the router as shown in Figure 6.2, in which the long distance link for nano communication can be realized, where finally each end user can perform in the same way of the hybrid transceiver. The storage tank is introduced by the sender (transmitter) to store the specimen that is required to use in the communication, whereas the required specimen is received and stored by the receiver end.

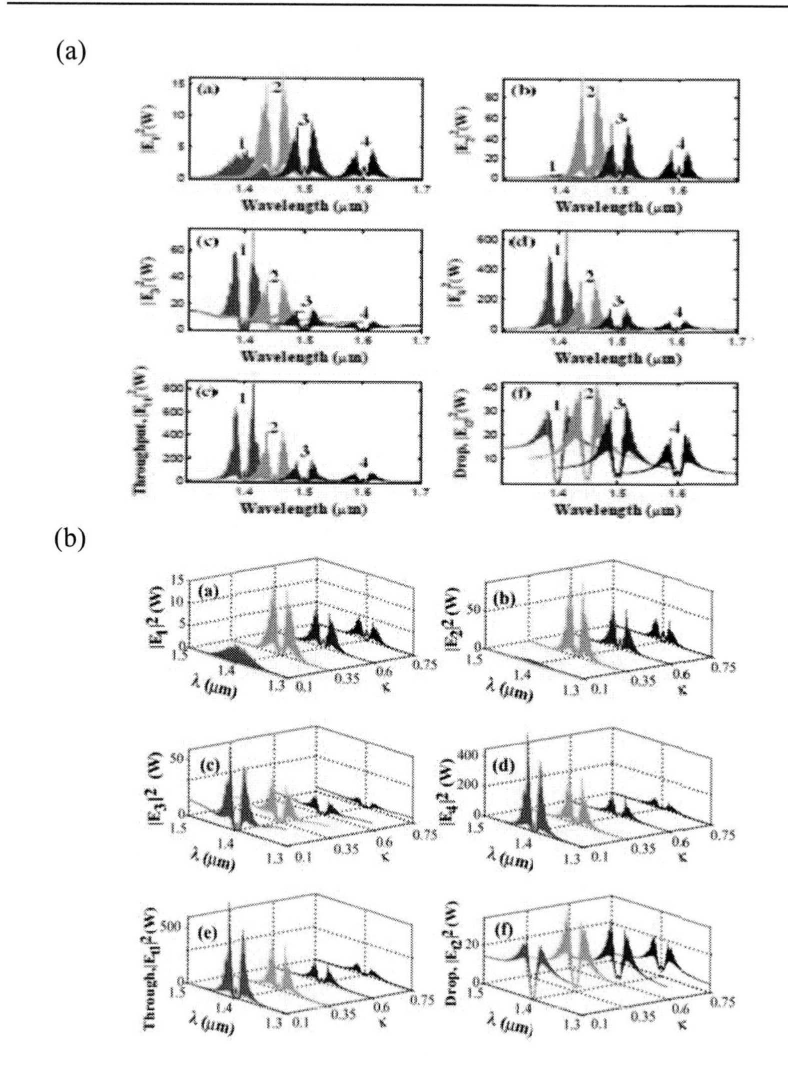

Figure 6.4. Simulation results, where (a) multi wavelength tweezers, (b) tunable tweezers.

6.3.2. Hybrid Repeater

In operation, the repeater is functioned in the way that there are two functions required for the repeater concept, where firstly, the repeater is performed when the number of microscopic volume is repeated as the initial transmitted by the transmitter. The second one is that the amplified gradient field is also obtained. In Figure 6.6, the repeater feature can be obtained by controlling the amount of the microscopic volume via the add port signal, in which the through and drop port signals in the form of gradient fields or trapping microscopic volume can be obtained.

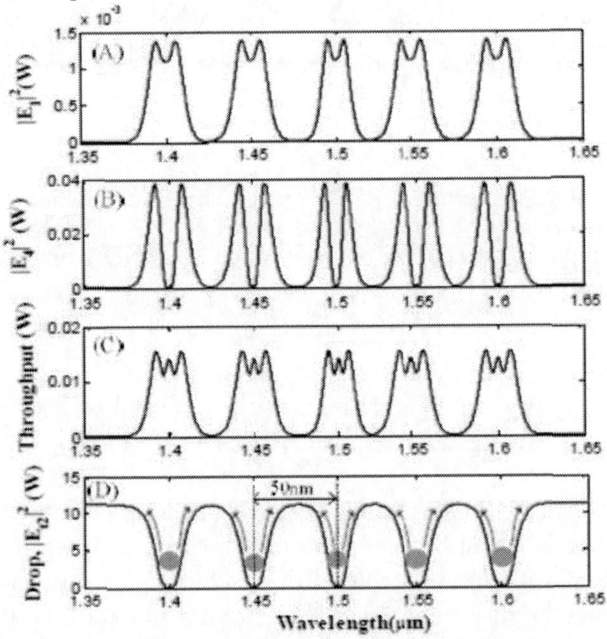

Figure 6.5. Simulation result of the dynamic tweezers with five different center wavelengths.

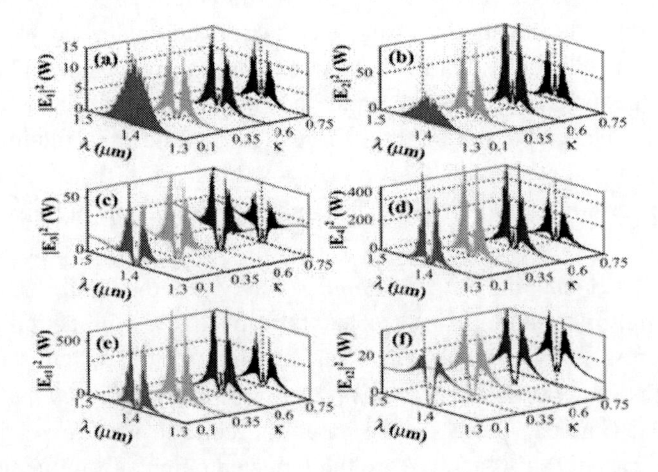

Figure 6.6. Repeater manipulation.

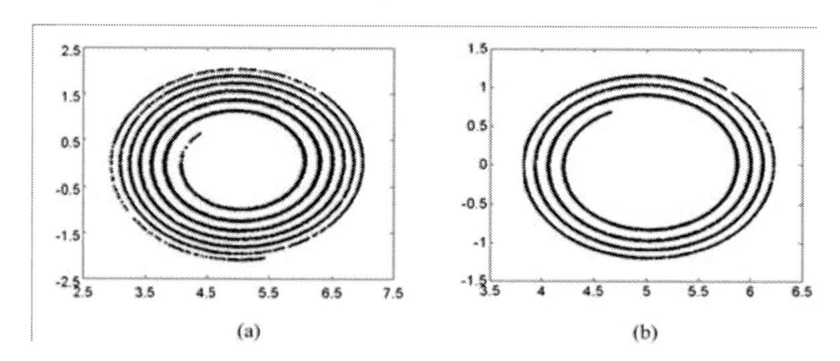

Figure 6.7. Spiral phase structure in PANDA ring (a) both within R_r and R_L (b) R_{ad}.

The spiral phase structure or optical vortice in PANDA ring are circulated spiral rotation. For the right and left microring within PANDA ring is fast rotated more than center of PANDA ring (R_{ad}) as shows in Figure 6.7..

6.4. Hybrid Interferometer

In figures 6.8(A) – 6.8(F), the tweezers in the form of potential wells are seen. These tweezers can be used for atom/molecule trapping. The potential well depth (peak valley) can be controlled by adjusting the system parameters, for instance, the bright soliton input power at the add port and the coupling coefficients. The potential well of the tweezers is tuned to be the multi wells and seen at the add port, as shown in figure 6.8(F). In application, the optical tweezers in the design system can be tuned and amplified as shown in figure 6.8. Therefore, the tunable optical tweezers can be controlled by the dark-bright soliton collision within the light signal multiplexer. This can be done by adjusting the parameters of the input power at the input and add ports, respectively. The output power at the through port is shown in figure 6.8(F).

The potential well with the optical power of *15 W* is observed. More results are as shown in figure 6.9. The coupling coefficients are given as $\kappa_0 = 0.1$, $\kappa_1 = 0.35$, $\kappa_2 = 0.1$ and $\kappa_3 = 0.2$, respectively. The ring radii are $R_{add} = 1 \mu m$, $R_{right} = 100nm$ and $R_{left} = 100nm$. A_{eff} are 0.50, 0.25 and $0.25\ \mu m^2$ [23] for add/drop, right and left ring resonators respectively. Simulation results of the ring resonator interferometer for atom/molecule spectroscopy with wavelengths centered at $\lambda_1 = 1400nm$, $\lambda_2 = 1450nm$, $\lambda_3 = 1500\ nm$. (A) dark solitons and Gaussian pulse are the input and control ports respectively, (B) $|E_2|^2$, (C) $|E_3|^2$, (D) $|E_4|^2$ and (E) are the reflected outputs from the throughput port, and (F) is the output at the drop port are shown in figure 6.9.

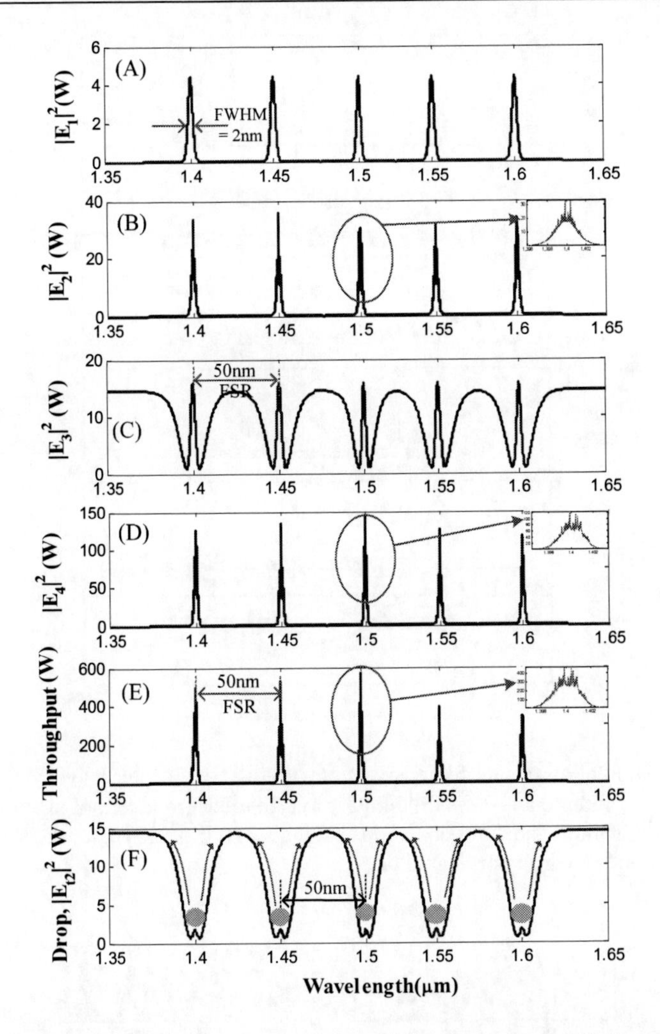

Figure 6.8. Results of dynamic tweezers array with atoms for five different center wavelengths, where (A) $|E_1|^2$, (B) $|E_2|^2$, (C) $|E_3|^2$, (D) $|E_4|^2$, (E) through port and (F) drop port signals.

In figure 6.10(A)-(F), to control the required well we adjust the coupling coefficients, which they are given as by $\kappa_0 = 0.2$, $\kappa_1 = 0.35$, $\kappa_2 = 0.1$ and $\kappa_3 = 0.1$, respectively, the simulation results are obtained for four different center wavelengths as shown, where (A) $|E_1|^2$, (B) $|E_2|^2$, (C) $|E_3|^2$ and (D) $|E_4|^2$ (E) are the through and (F) drop port signals. We found that the optical potential wells are stable and seen by using the suitable parameters in the add/drop design system.

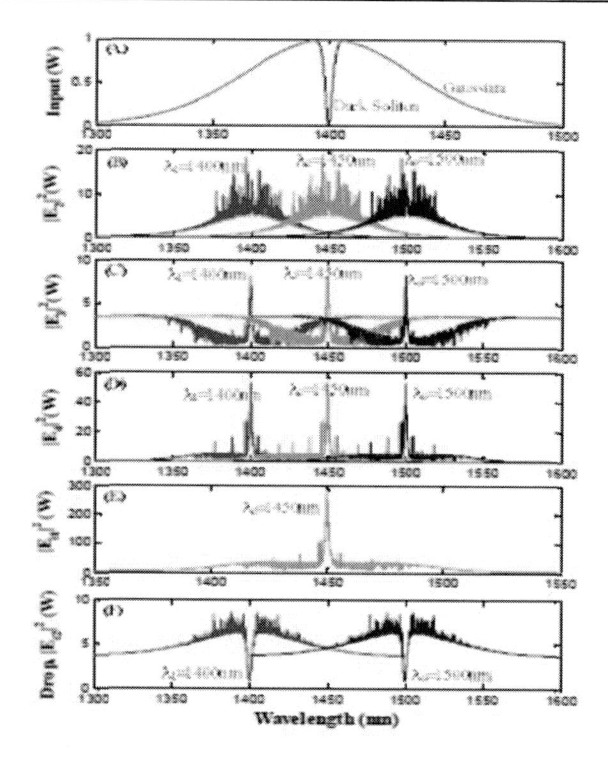

Figure 6.9. Simulation result of two tweezers with different center wavelengths λ_1 = 1400nm, λ_2 = 1450nm, and λ_3 = 1500nm for hybrid interferometer applications, where (A) input dark soliton and Gaussian control signals, (B) $|E_2|^2$, (C) $|E_3|^2$, (D) $|E_4|^2$, (E) through port and (F) drop port signals.

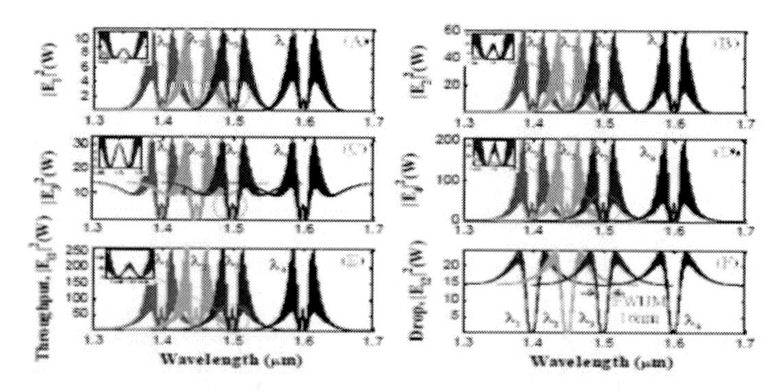

Figure 6.10. Simulation result of optical dynamic potential wells with four different center wavelengths , which they are controlled by using the Gaussian signal at the control port, where (A) $|E_1|^2$, (B) $|E_2|^2$, (C) $|E_3|^2$, (D) $|E_4|^2$, (E) through and (F) drop port signals, dark solitons and Gaussian pulse are the input and control ports respectively.

The obtained optical potential wells can be amplified and tuned as shown in figure 6.10(B) – (E), whereas the wells are seen. The double wells and single well outputs are obtained, whereas in this case the single well width of 16 nm is seen as shown in Figure 6.10(F).

Over the last three decades, interferometric measurement methods have been applied in research and industry for the investigation of deformation and vibration behaviour of mechanical components [24]. In applications, the term dynamics can be realized and is suitable for dynamic wells/tweezers control. This means that the movement of tweezers/well can be formed within the system. From the results, they have shown that the types of tweezers/wells can be configured whether they are in the stable/ or unstable configurations. For instance, the optical tweezers in the forms of peaks/valleys are kept in the stable forms within the add/drop filter. This can be seen at the through port in figures 6.3 and 6.4. One special case is observed in figure 4 when the double well is formed at E_3 as shown in figure 6.4(F). In operation, the optical signal multiplexer can be used to form a hybrid interferometer, signal calibration, hybrid clock signal generation, signal transmission error calculation and hybrid nano levitation mass method (LMM). In applications, the optical signal multiplexer can be used to form a hybrid interferometer, signal calibration, hybrid clock signal generation, signal transmission error calculation and hybrid nano levitation mass method (LMM), which can be described as follows:

(A)Hybrid interferometer: Atoms/molecules can be fed into the light signal multiplexer by dark soliton. The output atoms from the through port pass through the target and are reflected back to the light signal multiplexer. The induced change of the collision can be controlled by the add port input. Finally the interference signal is seen at the drop port.

(B)Signal error measurement and calibration: The transmission signal in any form of light signals can be input into the light signal multiplexer. The standard form of signal can be input into the add port. The comparative signals between the transmission and the standard signals can be seen at the drop port, which is then compared and formed the transmission error. Furthermore, the change in signal transition can be recovered by using the control parameter, which can be performed by the sensor, and the measurement can then be made.

(C)Hybrid clock signal generation: The hybrid clock can be formed by using the trapping atom, molecule, DNA, ion, electron to form the clock signals. The new clock can be used to form the new gate, which can be called a hybrid gate. By using the trapping atom, molecule, DNA, ion, electron, the gated operation can be formed which will be available for hybrid computer [25].

(D)Hybrid nano levitation mass method: Small charge particle or ion can be trapped and released from the through port to form the Doppler shift. This can be controlled and detected by using the control and drop ports respectively. Finally, the particle acceleration and force can be measured and calculated. Hence, by using the proposed method, the conventional LMM technique [26] can be manipulated by using the nano LMM.

6.5. CONCLUSION

We have shown that the multi photons (microscopic volumes) can be trapped by optical vortices (tweezers), which is generated a PANDA ring resonator. By utilizing the reasonable dark soliton input power, the dynamic multi photons can be controlled and stored (trapped) within the system. In this case, the dynamic optical vortices can be controlled and used to form the multi photons trapping tools. In application, such behaviors can be used to confine the suitable size of light pulse, atom or photon, which is then employed in the same way as the optical tweezers. But in this case the dynamic trapping photon/atom is realized, in which the trapped pulses or photons within the period of time (memory) within the system (PANDA ring) is plausible. Finally, we have shown that the use of a hybrid transceiver and repeater to form the long distance atom/molecule transportation will make the nano communication being realized in the near future. However, the problems of large atom/molecule and neutral atom/molecule may be occurred, in which the searching for new atom/molecule guide pipe and medium, for instance, nano tube and new medium will be the issue of investigation.

REFERENCES

[1] K. Sarapat, N. Sangwara, K. Srinuanjan, P. P Yupapin, and N. Pornsuwancharoen, "Novel dark-bright optical solitons conversion system and power amplification," *Opt. Eng.*, 48(4), 045004 (2009).

[2] T. Phatharaworamet, C. Teeka, R. Jomtarak, S. Mitatha, P. P. Yupapin, "Random binary code generation using dark-bright soliton conversion control within a PANDA ring resonator," *IEEE J. Lightw. Techn.*, 28(19), 2804 - 2809 2010.

[3] S. Mitatha, "Dark soliton behaviors within the nonlinear micro and nanoring resonators and applications," *Progress In Electromagnetic Research (PIER)*, 99, 383-404(2009).

[4] K. Kulsirirat, W. Techithdeera, P. P. Yupapin, "Dynamic potential well generation and control using double resonators incorporating in an add/drop filter," *Mod. Phys. Lett. B*, (2010), in press.

[5] T. Threepak, X. Louangvilay, S. Mitatha, and P. P. Yupapin, "Novel quantum-molecular transporter and networking via a wavelength router," *Microw. and Opt. Techn. Lett.*, 52, 1353-1357 (2010).

[6] M. Tasakorn, C. Teeka, R. Jomtarak, and P. P. Yupapin, "Multitweezers generation control within a nanoring resonator system," *Opt. Eng.*, 49, 075002 (2010).

[7] B. Piyatamrong, K. Kulsirirat, S. Mitatha, and P.P. Yupapin, "Dynamic potential well generation and control using double resonators incorporating in an add/drop filter," *Mod. Phys. Lett. B* 24(32): 3071–3082 (2010).

[8] M. A. Jalil, B. Piyatamrong, S. Mitatha, J. Ali, and P. P. Yupapin, "Molecular Transporter Generation for quantum-molecular transmission via an optical transmission line," *Nano Communication Network*, 2010. (in press)

[9] S. F. Hanim, J. Ali, and P. P. Yupapin, "Dark soliton generation using dual Brillouin fiber laser in a fiber optic ring resonator," *Microwave and Opt. Techn. Lett.,* 52, 881-883 (2010).

[10] P.P., Yupapin, T. Saktioto, and J. Ali, "Photon trapping model within a fiber Bragg grating for dynamic optical tweezers use," *Microwave and Opt. Techn. Lett.,* 52, 959-961 (2010).

[11] A.Ashkin, J. M. Dziedzic, J. E. Bjorkholm, and S. Chu, "Observation of a single-beam gradient force optical trap for dielectric particles," *Opt. Lett.,* 11, 288-290 (1986).

[12] K. Egashira, A. Terasaki, and T. Kondow, "Photon-trap spectroscopy applied to molecules adsorbed on a solid surface: probing with a standing wave versus a propagating wave," *Appl. Opt.*, 80, 5113-5115 (1998).

[13] A. V. Kachynski, A. N. Kuzmin, H. E. Pudavar, D. S. Kaputa, A. N. Cartwright, and P. N. Prasad, "Measurement of optical trapping forces

by use of the two-photon-excited fluorescence of microspheres," *Opt. Lett.,* 28, 2288-2290 (2003).

[14] M. Schulz, H. Crepaz, F. Schmidt-Kaler, J. Eschner, and R. Blatt, "Transfer of trapped atoms between two optical tweezer potentials," *J. Mod. Opt.,* 54, 1619-1626 (2007).

[15] A. Ashkin, "Forces of a single-beam gradient laser trap on a dielectric sphere in the ray optics regime," *J. Biophysics,* 61, 569-582 (1992).

[16] K. Svoboda, and S. M. Block, "Biological applications of optical forces," *Annual Review Biophysics and Bio-molecule Structure*, 247-282(1994).

[17] S. Mithata, N. Pornsuwancharoen, and P. P. Yupapin, "A simultaneous short wave and millimeter wave generation using a soliton pulse within a nano-waveguide," *IEEE J. Photon. Techn. Lett.*, 21(13), 932-934 (2009).

[18] Y. Kokubun, Y. Hatakeyama, M. Ogata, S. Suzuki, and N. Zaizen, "Fabrication technologies for vertically coupled microring resonator with multilevel crossing busline and ultracompact-ring radius," *IEEE J. Selec. Top. in Quan. Electron.,* 11, 4-10 (2005).

[19] P. P. Yupapin, and W. Suwancharoen, "Chaotic signal generation and cancellation using a micro ring resonator incorporating an optical add/drop multiplexer," *Opt. Comm*., 280, 343-350 (2007).

[20] P. P. Yupapin, P. Saeung, and C. Li, "Characteristics of complementary ring-resonator add/drop filters modeling by using graphical approach," *Opt. Comm.,* 272, 81-86 (2007).

[21] J. Zhu, S. K. Ozdemir, Y. F. Xiao, L. Li, L. He, D. R. Chen, and L. Yang, "On-chip single nanoparticle detection and sizing by mode splitting in an ultrahigh-Q microresonator," *Nat. Photon.*, **4**, 46-49 (2010).

[22] Q. Xu, D. Fattal, and R. G. Beausoleil, "Silicon microring resonators with 1.5-μm radius," *Opt. Express*, 16(6), 4309-4315 (2008).

[23] Kokubun Y, Hatakeyama Y, Ogata M, Suzuki S and Zaizen N "Fabrication technologies for vertically coupled microring resonator with multilevel crossing busline and ultracompact-ring radius," *IEEE J. Sel. Top. Quantum Electron.,* 11, 4-10 (2005).

[24] A. B. Matsko, "Practical Application of Micro resonators in Optics and Photonics" *CRC Press Taylor and Francis Group USA*, (2009).

[25] F. O'Mahony, C. P. Yue, M. A. Horowitz and S. S. Wong, "Design of a 10GHz clock distribution network using coupled standing-wave oscillators", *Proceedings of the 40th Conference on Design Automation*, 2003, pp. 682-687, XP002307842, Anahaim, CA.

[26] Y. Fujii and K. Maru, "Microforce materials tester," *Rev. Sci. Instrum.* 76, 065111, (2005).

Chapter 7

MOLECULAR COMMUNICATION

7.1. INTRODUCTION

Recently, Kulsirirat et al [1] have shown that by using the dark-bright soliton control within an add/drop filter, the dynamic potential well can be formed, which is available for molecule/atom trapping and transportation. The use of optical tweezers for molecule transportation within a wavelength router was also reported [2], where in this case the transport states are identified by using the correlated photons, where the entangled states of the molecular transporters were established. In general, optical tweezers are a powerful tool for use in the three-dimensional rotation of and translation (location manipulation) of nano-structures such as micro- and nano-particles as well as living micro-organisms [3]. Many research works have been concentrated on the static tweezers [4-8], which it cannot move. The benefit offered by optical tweezers is the ability to interact with nano-scaled objects in a non-invasive manner, i.e. there is no physical contact with the sample, thus preserving many important characteristics of the sample, such as the manipulation of a cell with no harm to the cell. Optical tweezers are now widely used and they are particularly powerful in the field of microbiology [9-11] to study cell–cell interactions, manipulate organelles without breaking the cell membrane and to measure adhesion forces between cells. In this paper we describe a new concept of developing an optical tweezers source using a dark soliton pulse. The developed tweezers have shown many potential applications in electron, ion, atom and molecule probing and manipulation as well as DNA probing and transportation. Furthermore, the soliton pulse generator is a simple and compact design, making it more commercially viable. In this paper, we present the theoretical background in the physical model concept, where potential well can be formed by the barrier of optical filed. The change in potential value, i.e. gradient of potential can produce force that can be used to confine/trap atoms/molecule. Furthermore, the change in potential well is still stable in some conditions, which mean that the dynamic optical tweezers is plausible, therefore, the transportation of atoms/molecules in the optical network via a dark soliton being realized in the near future. In application, the high capacity tweezers can be formed by using the tweezers array (multi tweezers)[12], which is available for high capacity transportation via optical wireless link [13]. The photon entanglement using a quantum processor is also reviewed.

7.2. Theoretical Background

Optical forces are customarily defined by the relationship [14]

$$F = \frac{Qn_m P}{c} \tag{7.1}$$

where Q is a dimensionless efficiency, n_m is the index of refraction of the suspending medium, c is the speed of light, and P is the incident laser power, measured at the specimen. Q represents the fraction of power utilized to exert force. For plane waves incident on a perfectly absorbing particle, $Q = 1$. To achieve stable trapping, the radiation pressure must create a stable, three-dimensional equilibrium. Because biological specimens are usually contained in aqueous medium, the dependence of F on nm can rarely be exploited to achieve higher trapping forces. Increasing the laser power is possible, but only over a limited range due to the possibility of optical damage. Q itself is therefore the main determinant of trapping force. It depends upon the NA, laser wavelength, light polarization state, laser mode structure, relative index of refraction, and geometry of the particle.

In the Rayleigh regime, trapping forces decompose naturally into two components. Since, in this limit, the electromagnetic field is uniform across the dielectric, particles can be treated as induced point dipoles. The scattering force is given by

$$F_{scatt} = n_m \frac{\langle S \rangle \sigma}{c}, \tag{7.2}$$

where

$$\sigma = \frac{8}{3}\pi (kr)^4 r^2 \left(\frac{m^2 - 1}{m^2 + 2} \right)^2 \tag{7.3}$$

is the scattering cross section of a Rayleigh sphere with radius r . S is the time-averaged Poynting vector, n is the index of refraction of the particle, $m = n / n_m$ is the relative index, and $k = 2\pi n_m / \lambda$ is the wave number of the light. Scattering force is proportional to the energy flux and points along the direction of propagation of the incident light. The gradient force is the Lorentz force acting on the dipole induced by the light field. It is given by

$$F_{grad} = \frac{\alpha}{2} \nabla \langle E^2 \rangle, \tag{7.4}$$

where

$$\alpha = n_m^2 r^3 \left(\frac{m^2 - 1}{m^2 + 2} \right) \tag{7.5}$$

is the polarizability of the particle. The gradient force is proportional and parallel to the gradient in energy density (for $rn > 1$). Stable trapping requires that the gradient force in the $-\hat{z}$ direction, against the direction of incident light, be greater than the scattering force. Increasing the NA decreases the focal spot size and increases the gradient strength [15].

To perform the proposed concept, a bright soliton pulse is introduced into the multi-stage nano ring resonators as shown in Figure 1, the input optical fields (E_{in}) of the bright and dark soliton pulses are given by an Eqs. (7.6) and (7.7) as [13]

$$E_{in}(t) = A\sec h\left[\frac{T}{T_0}\right]\exp\left[\left(\frac{z}{2L_D}\right) - i\omega_0 t\right] \tag{7.6}$$

$$E_{in}(t) = A\tan h\left[\frac{T}{T_0}\right]\exp\left[\left(\frac{z}{2L_D}\right) - i\omega_0 t\right] \tag{7.7}$$

where A and z are the optical field amplitude and propagation distance, respectively. *T* is a soliton pulse propagation time in a frame moving at the group velocity, T = t-β_1z, where β_1 and β_2 are the coefficients of the linear and second order terms of Taylor expansion of the propagation constant. $L_D = T_0^2/|\beta_2|$ is the dispersion length of the soliton pulse. T_0 in equation is the initial soliton pulse width. Where t is the soliton phase shift time, and he frequency shift of the soliton is ω_0. This solution describes a pulse that keeps its temporal width invariance as it propagates, and thus is called a temporal soliton. When a soliton peak intensity $(|\beta_2/\Gamma T_0^2|)$ is given, then T_0 is known. For the soliton pulse in the micro ring device, a balance should be achieved between the dispersion length (L_D) and the nonlinear length (L_{NL}= (1/ $\Gamma\phi_{NL}$), where Γ = n_2k_0, is the length scale over which dispersive or nonlinear effects makes the beam becomes wider or narrower. For a soliton pulse, there is a balance between dispersion and nonlinear lengths, hence $L_D = L_{NL}$.

We assume that the nonlinearity of the optical ring resonator is of the Kerr-type, i.e., the refractive index is given by

$$n = n_0 + n_2 I = n_0 + (\frac{n_2}{A_{eff}})P, \tag{7.8}$$

where n_0 and n_2 are the linear and nonlinear refractive indexes, respectively. I and P are the optical intensity and optical power, respectively. The effective mode core area of the device is given by A_{eff}. For the microring and nanoring resonators, the effective mode core areas range from 0.10 to 0.50 μm^2 [16].

When a Gaussian pulse is input and propagated within a fiber ring resonator, the resonant output is formed, thus, the normalized output of the light field is the ratio between the output and input fields ($E_{out}(t)$ and $E_{in}(t)$) in each roundtrip, which can be expressed as [15]

$$\left|\frac{E_{out}(t)}{E_{in}(t)}\right|^2 = (1-\gamma)\left[1 - \frac{(1-(1-\gamma)x^2)\kappa}{(1-x\sqrt{1-\gamma}\sqrt{1-\kappa})^2 + 4x\sqrt{1-\gamma}\sqrt{1-\kappa}\sin^2(\frac{\phi}{2})}\right] \tag{7.9}$$

Eq. (7.8) indicates that a ring resonator in the particular case is very similar to a Fabry-Perot cavity, which has an input and output mirror with a field reflectivity, (1-κ), and a fully reflecting mirror. k is the coupling coefficient, κ and $x = \exp(-\alpha L/2)$ represents a roundtrip loss coefficient, $\phi_0 = kLn_0$ and $\phi_{NL} = kL(\frac{n_2}{A_{eff}})P$ are the linear and nonlinear phase shifts, $k = 2\pi/\lambda$ is the wave propagation number in a vacuum. Where L and α are a waveguide length and linear absorption coefficient, respectively. In this work, the iterative method is introduced to obtain the results as shown in Eq. (7.9), similarly, when the output field is connected and input into the other ring resonators.

Figure 7.1 shows a schematic diagram of the multi variable quantum tweezers system, which two input fields at input and add ports. The dark soliton and bright soliton signals are input into input and add port respectively. The uplink carrier signal at 2 GHz is generated by using the dark soliton in a micro ring resonator system [13], where firstly the chaotic signal is generated with the rings R_1-R_4 , where the ring radii (R_1-R_7) are between 5- 12 μm, and the coupling coefficients are between 0.5- 0.98.

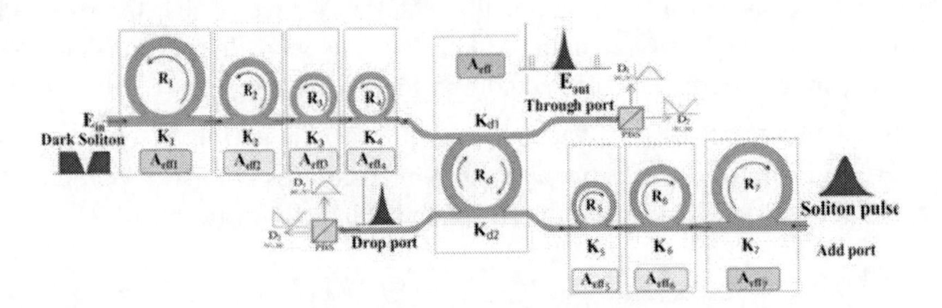

Figure 7.1. Shows a schematic diagram of the multi variable quantum tweezers generation system, where R_i : Ring radii, . K_s: Coupling coefficients, R_d: Add/Drop filter radius, K_{di}: Add/Drop coupling coefficients, A_{effs}: Effective core areas, $PBS_{:s}$: Polarizing Beamsplitters.

The input optical field as shown in Eq. (7.6), i.e. a Gaussian pulse, is input into a nonlinear microring resonator. By using the appropriate parameters, the chaotic signal is obtained by using Eq. (7.8). To retrieve the signals from the chaotic noise, we propose to use the add/drop device with the appropriate parameters. This is given in details as followings. The optical outputs of a ring resonator add/drop filter can be given by the Eqs. (7.10) and (7.11).

$$\left|\frac{E_t}{E_{in}}\right|^2 = \frac{(1-\kappa_1) - 2\sqrt{1-\kappa_1}\cdot\sqrt{1-\kappa_2}\,e^{-\frac{\alpha}{2}L}\cos(k_n L) + (1-\kappa_2)e^{-\alpha L}}{1+(1-\kappa_1)(1-\kappa_2)e^{-\alpha L} - 2\sqrt{1-\kappa_1}\cdot\sqrt{1-\kappa_2}\,e^{-\frac{\alpha}{2}L}\cos(k_n L)} \quad (7.10)$$

and

$$\left|\frac{E_d}{E_{in}}\right|^2 = \frac{\kappa_1\kappa_2 e^{-\frac{\alpha}{2}L}}{1+(1-\kappa_1)(1-\kappa_2)e^{-\alpha L} - 2\sqrt{1-\kappa_1}\cdot\sqrt{1-\kappa_2}\,e^{-\frac{\alpha}{2}L}\cos(k_n L)} \quad (7.11)$$

where E_t and E_d represents the optical fields of the throughput and drop ports respectively. The transmitted output can be controlled and obtained by choosing the suitable coupling ratio of the ring resonator, which is well derived and described by reference [17]. Where $\beta = kn_{eff}$ represents the propagation constant, n_{eff} is the effective refractive index of the waveguide, and the circumference of the ring is $L = 2\pi R$, here R is the radius of the ring. In the following, new parameters will be used for simplification, where $\phi = \beta L$ is the phase constant. The chaotic noise cancellation can be managed by using the specific parameters of the add/drop device, which the required signals at the specific wavelength band can be filtered and retrieved. $K_1$and K_2 are coupling coefficient of add/drop filters, $k_n = 2\pi/\lambda$ is the wave propagation number for in a vacuum, and the waveguide (ring resonator) loss is α = 0.5 dBmm^{-1}. The fractional coupler intensity loss is γ = 0.1. In the case of add/drop device [18], the nonlinear refractive index is neglected.

7.3. Multivariable Tweezers

Figure 7.2 shows the simulation results of (a) dark soliton input with 2 W peak power, (b) the chaotic signal is generated within the first ring (R_1), whereas the ring radius used is 12 μm, with the coupling coefficient (κ_1) is 0.9713, (c) the radius of the second ring (R_2) is 11.5 μm, with the coupling coefficient (κ_2) is 0.9723, (d) the radius of the third ring (R_3) is 10 μm, with the coupling coefficient (κ_3) is 0.9768 , the radius of the fourth ring (R_4) is 10 μm, and the coupling coefficient (κ_4) is 0.9768, (e) the input bright soliton with the center wavelength is at 1.500 μm ,(f) the drop port output signals(multi optical tweezers), (g) the throughput(through) port output signal amplitude and (h) the output signal at through port with the center wavelength is at 1.5 μm, and the radius of add/drop filter (R_d) is 50 μm, the coupling coefficient of add/drop filter is $\kappa_{d1}= \kappa_{d2}=0.5$, which the through port signal free spectrum rang is 0.002 μm.

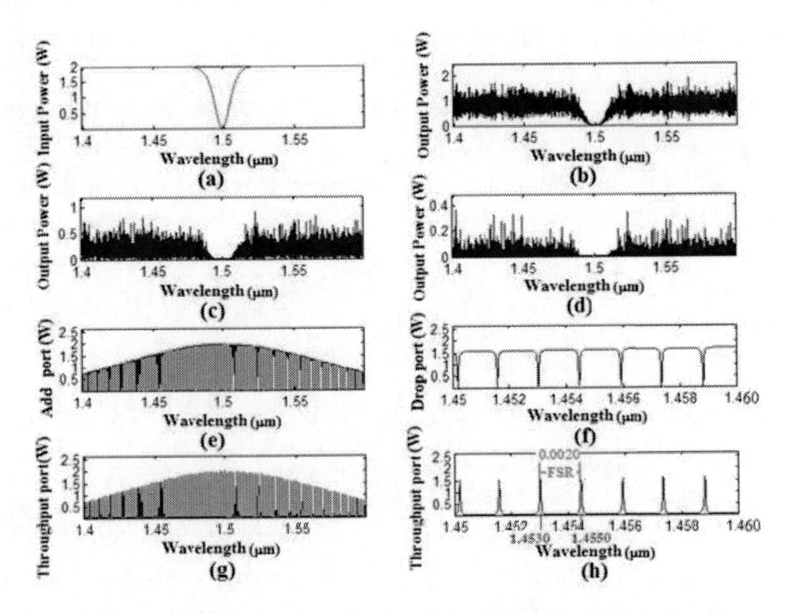

Figure 7.2. shows the input and output signals of the dynamic optical tweezers.

In Figure 7.3. a polarization coupler that separates the basic vertical and horizontal polarization states corresponds to an optical switch between the first and second incoming pulses. We assume those horizontally polarized pulses with a temporal separation of Δt. The coherence time of the consecutive pulses, i.e. between two pulses is larger than Δt. Then the following state is created by Eq. (7.12) [2, 19, 20].

$$|\Phi\rangle_p = |1,H\rangle_s |1,H\rangle_i + |2,H\rangle_s |2,H\rangle_i \tag{7.12}$$

In the expression $|k,H\rangle, k$ is the number of time slots (1 or 2), where | H> and | V> denotes the state of polarization, horizontal (H) or vertical (V), and the subscript identifies whether the state is the signal (s) or the idler (i) state. In Eq. (7.12), for simplicity we have omitted an amplitude term that is common to all product states. We employ the same simplification in subsequent equations in this paper. This two-photon state with H polarization shown by Eq. (7.12) is input into the orthogonal polarization-delay circuit shown schematically in Figure 7.3. The delay circuit consists of a coupler and the difference between the round-trip times of the micro ring resonator, which is equal to Δt . The micro ring is tilted by changing the round trip of the ring is converted into V at the delay circuit output. That is the delay circuits convert;

$$|k,H\rangle \text{ to } r|k,H\rangle + t_2 \exp(i\phi)|k+1,V\rangle + rt_2 \exp(i_2\phi)|k+2,H\rangle + r_2 t_2 \exp(i_3\phi)|k+3,V\rangle \tag{7.13}$$

where t and r is the amplitude transmittances to cross and bar ports in a coupler. Then Eq. (7.8) is converted into the polarized state by the delay circuit as

$$\begin{aligned}|\Phi\rangle &= [|1,H\rangle_s + \exp(i\phi_s)|2,V\rangle_s] \times [|1,H\rangle_i + \exp(i\phi_i)|2,V\rangle_i] + [|2,H\rangle_s + \exp(i\phi_s)|3,V\rangle_s] \\ &\quad \times [|2,H\rangle_i + \exp(i\phi_i)|2,V\rangle_i] \\ &= [|1,H\rangle_s|1,H\rangle_i + \exp(i\phi_i)|1,H\rangle_s|2,V\rangle_i] + \exp(i\phi_s)|2,V\rangle_s|1,H\rangle_i \\ &\quad + \exp[i(\phi_s + \phi_i)]|2,V\rangle_s|2,V\rangle_i + |2,H\rangle_s|2,H\rangle_i + \exp(i\phi_i|2,H\rangle_s|3,V\rangle_i \\ &\quad + \exp(i\phi_s|3,V\rangle_s|2,H\rangle_i + \exp[i(\phi_s + \phi_i)]|3,V\rangle_s|3,V\rangle_i\end{aligned} \tag{7.14}$$

By the coincidence counts in the second time slot, we can extract the fourth and fifth terms. As a result, we can obtain the following polarization entangled state as

$$|\Phi\rangle = |2,H\rangle_s|2,H\rangle_i + \exp[i(\phi_s + \phi_i)]|2,V\rangle_s|2,V\rangle_i \tag{7.15}$$

We assume that the response time of the Kerr effect is much less than the cavity round-trip time. Because of the Kerr nonlinearity of the optical fiber, the strong pulses acquire an intensity dependent phase shift during propagation. The interference of light pulses at a coupler introduces the out put beam, which is entangled. Due to the polarization states of light pulses are changed and converted while circulating in the delay circuit, where the polarization entangled photon pairs can be generated. The entangled photons of the nonlinear ring resonator are separated to be the signal and idler photon probability. The polarization angle adjustment device as shown in Figure 7.3 is applied to investigate the orientation and optical output intensity.

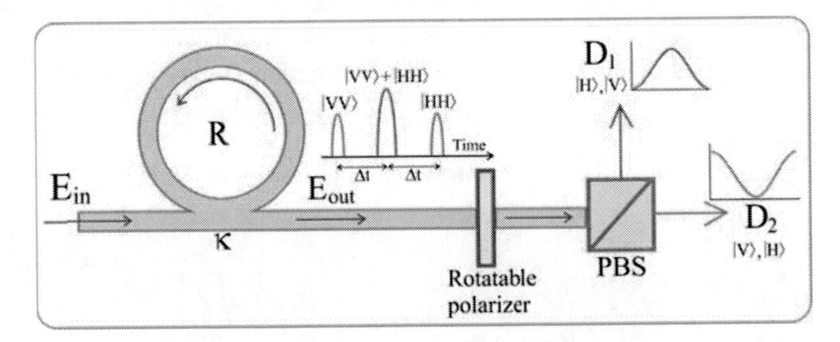

Figure 7.3. A schematic of the entangled photon generation system; where PBS: Polarizing Beam Splitter, Ds: Detectors.

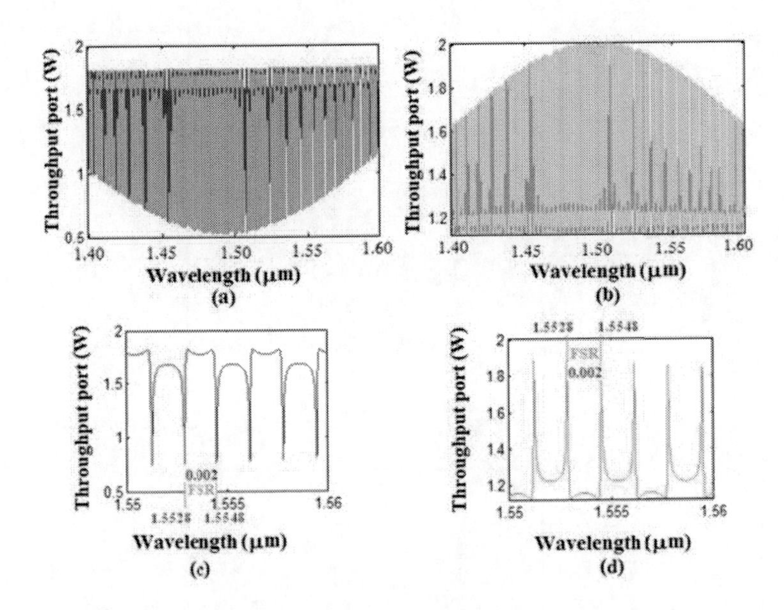

Figure 7.4. Results of the multi variable quantum tweezers.

In operation, a dark soliton input power of 2 W is input into the system as shown in Figure 7.1. The first ring radius (R_1) is 12 μm, and the coupling coefficient (κ_1) is 0.9713. The radius of the second ring (R_2) is 11.5 μm, where the coupling coefficient (κ_2) is 0.9723. The radius of the third ring (R_3) is 10 μm, the coupling coefficient (κ_3) is 0.9768, where the radius of the fourth ring

(R_4) is 10 μm, and the coupling coefficient (κ_4) is 0.9768. The bright soliton with the center wavelength at 1.500 μm is input into the add port. The coupling coefficient of add/drop filter is $\kappa_{d1}= \kappa_{d2}=0.5$. Figure 4 shows results of the multi quantum tweezers using the entangled photon states identification, where (a) is $|E^2||V\rangle$, which is represented the intensity of vertical polarization, where (b) $|E^2||H\rangle$ is the intensity in the horizontal polarization. Where (c) and (d) is the multi quantum tweezers within the wavelength range between 1.5528 to 1.5548 μm, whereas the free spectrum range is 0.002 μm within the period of 100 ns.

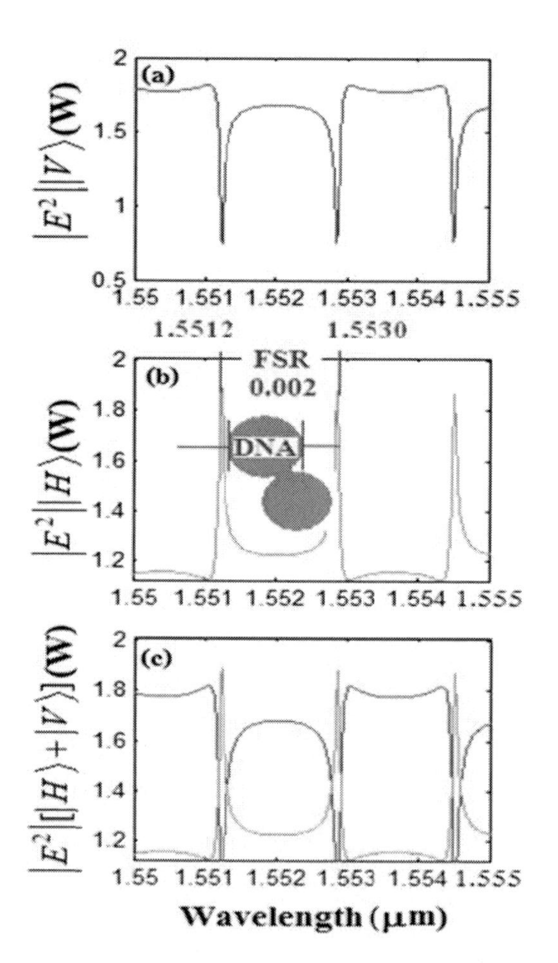

Figure 7.5. Shows the multi variable quantum tweezers with trapped DNAs.

Figure 7.5. shows the results of the dynamic quantum tweezers (wells), where the tweezers intensity at the throughput port is (a) $|E^2||V\rangle$, which is the vertical polarization intensity, (b) $|E^2||H\rangle$ is the horizontal polarization intensity. For instance, these can be used to form the quantum tweezers for DNA size of 0.34 nm, and (c) is $|E^2|[|H\rangle+|V\rangle]$ is the vertical $|V\rangle$ and horizontal $|H\rangle$ polarization intensity respectively for dynamic quantum tweezers, which are performed by using the entangled photons as shown in Figs. 1 and 5.

7.4. CONCLUSION

We have shown that the multi variable quantum tweezers array can be generation by using a microring resonator system for use of quantum-molecular and atomic tweezers transmission, whereas the free spectrum range obtained is between 2.0-5.0 nm, which is shown the quantum tweezers array function. In applications, the smallest and simple quantum tweezes system can be employed, where the quantum states, entangle photon and multi tweezers can be generated, controlled and trapping , then transmitted via the transmission lines, which is allowed to use for multi molecules or atoms transportation via the optical wireless link. Finally, we have claimed this system is the novel design for multi variable quantum tweezers array by using dark-bright soliton control within an add/drop multiplexer, which is a possible technique that can be form the DNA with quantum states identification.

REFERENCES

[1] B. Piyatamrong, K. Kulsirirat, W. Techithdeera, S. Mitatha, P.P. Yupapin, "Dynamic potential well generation and control using double resonators incorporating in an add/drop filter. *Mod Phys Lett B.* 2010;24:3071–3082

[2] T. Threepak, X. Luangvilay, S. Mitatha and P.P. Yupapin, "Novel quantum-molecular transporter and networking via a wavelength router," *Microw. and Opt. Technol. Lett.,* 52, 1353-1357(2010).

[3] A. Ashkin, J.M. Dziedzic, J.E. Bjorkholm and S. Chu, "Observation of a single-beam gradient force optical trap for dielectric particles," *Opt. Lett.,* 11, 288-290(1986).

[4] R. L. Eriksen, V. R. Daria, and J. Gluckstad, "Fully dynamic multiple-beam optical tweezers," *Opt. Express*. 10, 597–602 (2002).

[5] P. J. Rodrigo, V. R. Daria, and J. Gluckstad, "Real-time interactive optical micromanipulation of a mixture of high- and low-index particles," *Opt. Express,* 12, 1417–1425 (2004),

[6] J. Liesener, M. Reicherter, T. Haist, and H. J. Tiziani, "Multi-functional optical tweezers using computer generated holograms," *Opt. Commun.* 185, 77–82 (2000).

[7] J. E. Curtis, B. A. Koss, and D. G. Grier, "Dynamic holographic optical tweezers," *Opt. Commun*. 207, 169–175 (2002).

[8] W. J. Hossack, E. Theofanidou, J. Crain, K. Heggarty, and M. Birch, "High-speed holographic optical tweezers using a ferroelectric liquid crystal microdisplay," *Opt. Express* 11, 2053–2059 (2003),

[9] V. Boyer, R. M. Godun, G. Smirne, D. Cassettari, C. M.Chandrashekar, A. B. Deb, Z. J. Laczik and C. J. Foot, "Dynamic manipulation of Bose-Einstein condensates with a spatial light modulator", *Phys. Rev., A* 73, 031402(R) (2006).

[10] A.V. Carpentier, J. Belmonte-Beitia, H. Michinel and V.M. Perez-Garcia, "Laser tweezers for atomic solitons," *J. of Mod. Opt*., 55(17), 2819–2829(2008).

[11] V. Milner, J. L. Hanssen, W. C. Campbell and M. G. Raizen, "Optical billiards for atoms, *Phys. Rev. Lett*., 86, 1514-1516(2001).

[12] P. T. Korda, M. B. Taylor, and D. G. Grier, "Kinetically locked-in colloidal transport in an array of optical tweezers," *Phys. Rev. Lett.* 89, 128301 (2002).

[13] S. Mithata, N. Pornsuwancharoen and P.P. Yupapin, "A simultaneous short wave and millimeter wave generation using a soliton pulse within a nano-waveguide," *IEEE Photon. Technol. Lett.,* 21, 932-934(2009).

[14] A. Ashkin . Forces of a single-beam gradient laser trap on a dielectric sphere in the ray optics regime. *Biophys. J.* 61:569-82(1992).

[15] K. Svoboda and S. M. Block, " Biological Applications of Optical Forces" *Annu. Rev. Biophs. Biomol. Struct*.,247-282(1994).

[16] Y. Kokubun, Y. Hatakeyama, M. Ogata, S. Suzuki, and N. Zaizen, "Fabrication technologies for vertically coupled micro ring resonator with multilevel crossing busline and ultracompact-ring radius", *IEEE J. Sel. Top. Quantum Electron.*, 11,4–10(2005).

[17] P.P. Yupapin and W. Suwancharoen, "Chaotic signal generation and cancellation using a microring resonator incorporating an optical add/drop multiplexer," *Opt. Commun*., 280, 343-350(2007).

[18] P.P. Yupapin, P. Saeung and C. Li, "Characteristics of complementary ring-resonator add/drop filters modeling by using graphical approach", *Opt. Commun.*, 272, 81-86 (2007).

[19] P.P. Yupapin, "Generalized quantum key distribution via micro ring resonator for mobile telephone networks", *Int. J. Light and Electron Opt.,* (2008)DOI : 10.1016/j.ijleo.2008.07.030.

[20] C.H. Bennett, G. Brassard, C. Crepeau, R. Jozsa, A. Peres, and W.K. Wootters, "Teleporting an unknown quantum state via dual classical an Eistein-Podolsky-Rosen," *Phys. Rev. Lett.*,70,1895-1899(1993).

Chapter 8

MOLECULAR NETWORKS

8.1. INTRODUCTION

Optical tweezers provide an excellent mechanism to precisely measure and manipulate objects as small as a single atom and have been recognized as a powerful tool for the use in many researches [1-7]. For example, the use for three-dimensional rotation and translation (location manipulation) of nano-structures such as micro- and nano-particles as well as living micro-organisms has been reported [5, 6]. The benefit offered by optical tweezers is the ability to interact with nano-scaled objects in a non-invasive manner, i.e. there is no physical contact (non-demolition) with the sample, thus preserving many important characteristics of the sample, such as the manipulation of a cell with no harm to the cell. Optical tweezers are now widely used and they are particularly powerful in the field of microbiology to study cell–cell interactions, manipulate organelles without breaking the cell membrane and to measure adhesion forces between cells. In this paper we describe a new concept of developing an optical tweezers source using a Gaussian pulse (laser pulse) to use in atomic spectroscopy. The developed tweezes have many potential applications in electron, ion, atom and molecule probing and manipulation as well as DNA probing, transportation and imaging. We present the theoretical background in the physical model concept, where a potential well can be formed by the barrier of an optical field, i.e. laser pulses. Also, the experimental of optical tweezers have been tested to verify the result of optical tweezers generated by using a Gaussian pulse in a fiber optic amplification system. The advantage of this solution is that the optical tweezers acts as a moveable and tunable probe, which can be used in any location of the interesting target, which the resolution of the atom size is obtained.

Recently, several research works have shown that use of dark and bright soliton in various applications can be realized [8-13], where one of them has shown that the secured signals in the communication link can be retrieved by using a suitable add/drop filter that is connected to the transmission line. Another promising application of a dark soliton signal [14] is for the large guard band of two different frequencies which can be achieved by using a dark soliton generation scheme and trapping a dark soliton pulse within a nanoring resonator [15]. Furthermore, the dark soliton pulse shows a more stable behavior than the bright solitons with respect to the perturbations such as amplifier noise, fiber losses, and intra-pulse stimulated Raman scattering [16]. It is found that the dark soliton pulses propagation in a lossy fiber, spreads in time at approximately half the rate of bright solitons. Recently, the localized dark solitons in the add/drop filter has been reported [17]. In this paper, the use of dark and bright solitons propagating within the proposed ring resonator systems is investigated and described, where the use of suitable parameters based on the realistic device is discussed. The potential of using generated dark soliton signals for single photon tweezers and molecular transporter, especially for the hybrid quantum-molecular communication and transportation in the communication network, is described in detail.

8.2. Transporter Generation

Bright and dark soliton pulses are introduced into the multi-stage nanoring resonators as shown in Figure 8.1. The input time dependent optical field (E_{in}) of the bright and dark soliton pulses input are given by an Eq. (8.1) and (8.2)[15], respectively.

$$E_{in}(t) = A \sec h\left[\frac{T}{T_0}\right] \exp\left[\left(\frac{z}{2L_D}\right) - i\omega_0 t\right] \tag{8.1}$$

and

$$E_{in}(t) = A \tanh\left[\frac{T}{T_0}\right] \exp\left[\left(\frac{z}{2L_D}\right) - i\omega_0 t\right] \tag{8.2}$$

where A and z are the optical field amplitude and propagation distance, respectively. T is a soliton pulse propagation time in a frame moving at the group velocity, $T=t-\beta_1*z$, where β_1 and β_2 are the coefficients of the linear and second-order terms of Taylor expansion of the propagation constant. $L_D= T_0^2/|\beta_2|$ is the dispersion length of the soliton pulse. T_0 in equation is a soliton pulse propagation time at initial input (or soliton pulse width), where t is the soliton phase shift time, and the frequency shift of the soliton is ω_0. This solution describes a pulse that keeps its temporal width invariance as it propagates, and thus is called a temporal soliton. When a soliton peak intensity $(|\beta_2/\Gamma T_0^2|)$ is given, then T_0 is known. For the soliton pulse in the microring device, a balance should be achieved between the dispersion length (L_D) and the nonlinear length ($L_{NL}=1/\Gamma\phi_{NL}$), where $\Gamma=n_2*k_0$, is the length scale over which dispersive or nonlinear effects makes the beam become wider or narrower. For a soliton pulse, there is a balance between dispersion and nonlinear lengths, hence $L_D = L_{NL}$. Similarly, the output soliton of the system in Figure 8.2 can be calculated by using Gaussian equations as given in the above case.

We assume that the nonlinearity of the optical ring resonator is of the Kerr-type, i.e., the refractive index is given by

$$n = n_0 + n_2 I = n_0 + (\frac{n_2}{A_{eff}})P, \tag{8.3}$$

where n_0 and n_2 are the linear and nonlinear refractive indexes, respectively. I and P are the optical intensity and optical power, respectively. The effective mode core area of the device is given by A_{eff}. For the microring and nanoring resonators, the effective mode core areas range from 0.10 to 0.50 μm^2 [18, 19]

When a Gaussian pulse is input and propagated within a fiber ring resonator, the resonant output is formed, thus, the normalized output of the light field is the ratio between the output and input fields ($E_{out}(t)$ and $E_{in}(t)$) in each roundtrip, which can be expressed as [20]

$$\left|\frac{E_{out}(t)}{E_{in}(t)}\right|^2 = (1-\gamma)\left[1-\frac{(1-(1-\gamma)x^2)\kappa}{(1-x\sqrt{1-\gamma}\sqrt{1-\kappa})^2+4x\sqrt{1-\gamma}\sqrt{1-\kappa}\sin^2(\frac{\phi}{2})}\right] \tag{8.4}$$

Equation (8.4) indicates that a ring resonator in the particular case is very similar to a Fabry-Perot cavity, which has an input and output mirror with a field reflectivity, (1-κ), and a fully reflecting mirror. k is the coupling coefficient, κ and $x = \exp(-\alpha L/2)$ represents a roundtrip loss coefficient, $\phi_0 = kLn_0$ and $\phi_{NL} = kL(\frac{n_2}{A_{eff}})P$ are the linear and nonlinear phase shifts, $k = 2\pi/\lambda$ number in a vacuum, where L and are a waveguide length and linear absorption coefficient, respectively. In this work, the iterative method is introduced to obtain the results as shown in equation (8.4), similarly, when the output field is connected and input into the other ring resonators.

The input optical field as shown in equations (8.1) and (8.2), i.e. a soliton pulse, is input into a nonlinear microring resonator. By using the appropriate parameters, the chaotic signal is obtained by using equation (8.4). To retrieve the signals from the chaotic noise, we propose to use the add/drop device with the appropriate parameters. This is given in details as followings. The optical outputs of a ring resonator add/drop filter can be given by the equations (8.5) and (8.6)[21].

$$\left|\frac{E_t}{E_{in}}\right|^2 = \frac{(1-\kappa_1) - 2\sqrt{1-\kappa_1}\cdot\sqrt{1-\kappa_2}\,e^{-\frac{\alpha}{2}L}\cos(k_n L) + (1-\kappa_2)e^{-\alpha L}}{1 + (1-\kappa_1)(1-\kappa_2)e^{-\alpha L} - 2\sqrt{1-\kappa_1}\cdot\sqrt{1-\kappa_2}\,e^{-\frac{\alpha}{2}L}\cos(k_n L)} \tag{8.5}$$

and

$$\left|\frac{E_d}{E_{in}}\right|^2 = \frac{\kappa_1\kappa_2 e^{-\frac{\alpha}{2}L}}{1 + (1-\kappa_1)(1-\kappa_2)e^{-\alpha L} - 2\sqrt{1-\kappa_1}\cdot\sqrt{1-\kappa_2}\,e^{-\frac{\alpha}{2}L}\cos(k_n L)} \tag{8.6}$$

where E_t and E_d represents the optical fields of the throughput and drop ports respectively. The transmitted output can be controlled and obtained by choosing the suitable coupling ratio of the ring resonator, which is well derived and described by reference [21]. Where n_{eff} is the effective refractive index of the waveguide, and the circumference of the ring is $L = 2\pi R$, here R is the radius of the ring. In the following, new parameters will be used for simplification, $\phi = \beta L$ is the phase constant, where β is the propagation constant. The chaotic noise cancellation can be managed by using the specific parameters of the add/drop device, which the required signals at the specific wavelength band can be filtered and retrieved. $K_1$and K_2 are coupling coefficient of add/drop filters, $k_n = 2\pi / \lambda$ is the wave propagation number for in a vacuum, and the waveguide (ring resonator) loss is α = 0.5 $dBmm^{-1}$. The fractional coupler intensity loss is γ = 0.1. In the case of add/drop device, the nonlinear refractive index is neglected.

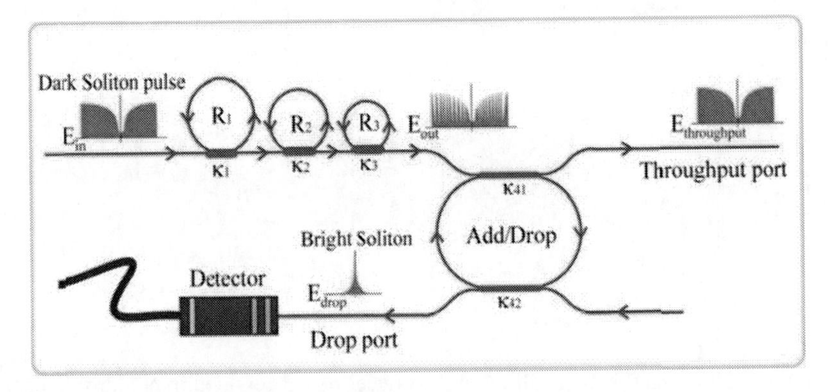

Figure 8.1. Schematic of a dark-bright soliton conversion system, where R_s is the ring radii, κ_s is the coupling coefficient, and κ_{41} and κ_{42} are the add/drop coupling coefficients.

In operation, a dark soliton pulse with 50-ns pulse width with the maximum power of 0.65W is input into the dark-bright soliton conversion system as shown in Figure 8.1. The suitable ring parameters are ring radii, where R_1=10.0μm, R_2=7.0μm, and R_3=5.0μm. In order to make the system associate with the practical device [18, 20] the selected parameters of the system are fixed to λ_0=1.50μm, n_0=3.34 (InGaAsP/InP). The effective core areas are A_{eff}=0.50, 0.25, and 0.10 μm^2 for a microring resonator (MRR) and nanoring resonator (NRR), respectively. The waveguide and coupling loses are α =0.5$dBmm^{-1}$ and γ =0.1, respectively, and the coupling coefficients κ_s of the MRR are ranged from 0.05 to 0.90. The nonlinear refractive index is $n_2=2.2\times10^{-13}$ m^2/W. In this case, the waveguide loss used is 0.5 $dBmm^{-1}$. The input dark soliton pulse is chopped (sliced) into the smaller output signals of the filtering signals within the rings R_2 and R_3. We find that the output signals from R_3 are smaller than from R_1, which is more difficult to detect when it is used in the link. In fact, the multistage ring system is proposed due to the different core effective areas of the rings in the system, where the effective areas can be transferred from 0.50 to 0.10μm^2 with some losses.

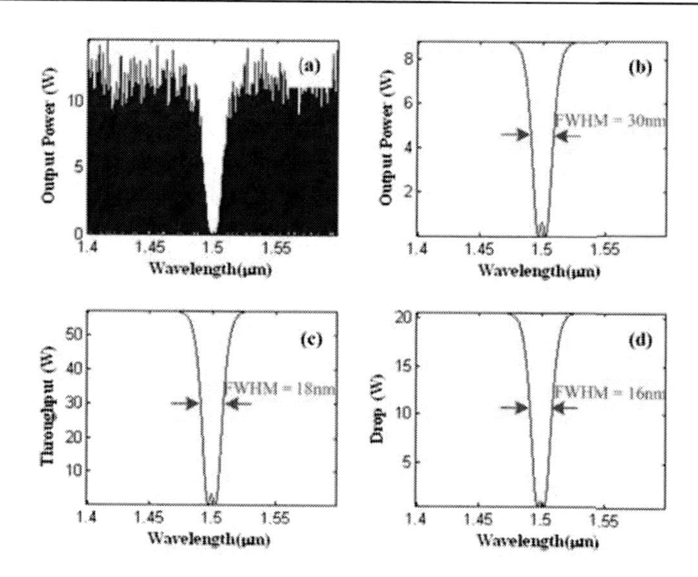

Figure 8.2. The dynamic dark soliton(optical tweezers) occurs within add/drop tunable filter, where (a) add/drop signals, (b) dark – bright soliton collision, (c) optical tweezers at throughput port, and (d) optical tweezers at drop port.

The soliton signals in R_3 is entered in the add/drop filter, where the dark-bright soliton conversion can be performed by using Eqs. (8.5) and (8.6). The dynamic dark soliton control can be configured to be an optical dynamic tool known as an optical tweezers, where more details of optical tweezers can be found in references [22, 23]. After the bright soliton input is added into the system via add port as shown in Figure 8.2, the optical tweezers behavior is observed. The parameters of system are used the same as the previous case. A bright soliton is generated with the central wavelength l0 = 1.5μm. When the bright soliton propagates into the add/drop system, a dark-bright soliton collision in the add/drop system is observed. The dark valley dept, i.e. potential well is changed when it is modulated by the trapping energy (dark-bright solitons interaction) as shown in Figure 8.2. The dynamic dark soliton(optical tweezers) occurs within add/drop tunable filter, when the bright soliton is input into the add port with the central wavelength λ_0 = 1.5μm. (a) add/drop signals, (b) dark – bright soliton collision, (c) optical tweezers at throughput port, and (d) optical tweezers at drop port. The recovery photon can be obtained by using the dark-bright soliton conversion, which is well analyzed by Sarapat et al. [13]. The trapped photon or molecule can be released or separated from the dark soliton pulse and, in this case, the bright soliton becomes alive and can be seen.

8.3. TRANSPORTER QUANTUM STATE

Let us consider the case when the optical tweezers output from the throughput port in Figs. 8.1 and 8.2 is partially input into the quantum processor unit as shown in Figs. 8.3 and 8.4. Generally, there are two pairs of possible polarization entangled photons forming within the ring device, which are represented by the four polarization orientation angles as [0°, 90°], [135° and 180°]. These can be formed by using the optical component called the polarization rotatable device and a polarizing beam splitter (PBS). In this concept, we assume that the polarized photon can be performed by using the proposed arrangement where each pair of the transmitted qubits can randomly form the entangled photon pairs. To begin this concept, we introduce the technique that can be used to create the entangled photon pair (qubits) as shown in Figure 8.4, a polarization coupler that separates the basic vertical and horizontal polarization states corresponds to an optical switch between the short and the long pulses. We assume those horizontally polarized pulses with a temporal separation of Δt. The coherence time of the consecutive pulses is larger than Δt. Then the following state is created by Eq. (8.7) [24].

$$|\Phi\rangle_p = |1,H\rangle_s |1,H\rangle_i + |2,H\rangle_s |2,H\rangle_i \tag{8.7}$$

In the expression $|k,H\rangle$, k is the number of time slots (1 or 2), where denotes the state of polarization [horizontal $|H\rangle$or vertical $|V\rangle$], and the subscript identifies whether the state is the signal (s) or the idler (i) state. In Eq. (8.7), for simplicity, we have omitted an amplitude term that is common to all product states. We employ the same simplification in subsequent equations in this paper. This two-photon state with $|H\rangle$polarization shown by Eq. (8.7) is input into the orthogonal polarization-delay circuit shown schematically. The delay circuit consists of a coupler and the difference between the round-trip times of the microring resonator, which is equal to Δt. The microring is tilted by changing the round trip of the ring is converted into $|V\rangle$at the delay circuit output. That is the delay circuits convert $|k,H\rangle$to be

$$r|k,H\rangle + t_2 \exp(i\phi)|k+1,V\rangle + rt_2 \exp(i_2\phi)|k+2,H\rangle + r_2 t_2 \exp(i_3\phi)|k+3,V\rangle$$

where t and r is the amplitude transmittances to cross and bar ports in a coupler. Then Eq. (8.7) is converted into the polarized state by the delay circuit as

$$\begin{aligned}|\Phi\rangle &= \left[|1,H\rangle_s + \exp(i\phi_s)|2,V\rangle_s\right]\left[|1,H\rangle_i + \exp(i\phi_i)|2,V\rangle_i\right] \\ &\quad + \left[|2,H\rangle_s + \exp(i\phi_s)|3,V\rangle_s\right]\left[|2,H\rangle_i + \exp(i\phi_i)|3,V\rangle_i\right] \\ &= \left[|1,H\rangle_s|1,H\rangle_i + \exp(i\phi_s)|1,H\rangle_s|2,V\rangle_i\right] \\ &\quad + \exp(i\phi_s)|2,V\rangle_s|1,H\rangle_i + \exp(i(\phi_s+\phi_i))|2,V\rangle_s|2,V\rangle_i \\ &\quad + |2,H\rangle_s|2,H\rangle_i + \exp(i\phi_i)|2,H\rangle_i|3,V\rangle_i \\ &\quad + \exp(i\phi_s)|3,V\rangle_s|2,H\rangle_i + \exp(i(\phi_s+\phi_i))|3,V\rangle_s|3,V\rangle_i\end{aligned} \tag{8.8}$$

By the coincidence counts in the second time slot, we can extract the fourth and fifth terms. As a result, we can obtain the following polarization entangled state as

$$|\Phi\rangle = |2,H\rangle_s|2,H\rangle_i + \exp[i(\phi_s+\phi_i)]|2,V\rangle_s|2,V\rangle_i \tag{8.9}$$

We assume that the response time of the Kerr effect is much less than the cavity round-trip time. Because of the Kerr nonlinearity of the optical device, the strong pulses acquire an intensity dependent phase shift during propagation. The interference of light pulses at a coupler introduces the output beam, which is entangled. Due to the polarization states of light pulses are changed and converted while circulating in the delay circuit, where the polarization entangled photon pairs can be generated. The entangled photons of the nonlinear ring resonator are separated to be the signal and idler photon probability. The polarization angle adjustment device is applied to investigate the orientation and optical output intensity and is well described by the published work [24, 25]. The transporter states can be controlled and identified by using the quantum processing system as shown in Figs. 8.3 and 8.4.

8.4. Multi-Quantum Molecular Transportation

By using the reasonable dark-bright soliton input power, the tunable optical tweezers can be controlled, which can provide the entangled photon as the dynamic optical tweezers probe. The smallest tweezer width of 16 nm is generated and achieved. In application, such a behavior can be used to confine the suitable size of light pulse or molecule, which can be employed in the same way of the optical tweezers. But in this case the terms dynamic probing is come to be a realistic function, therefore, the transportation of the trapped atom/molecule/photon by a single photon is plausible.

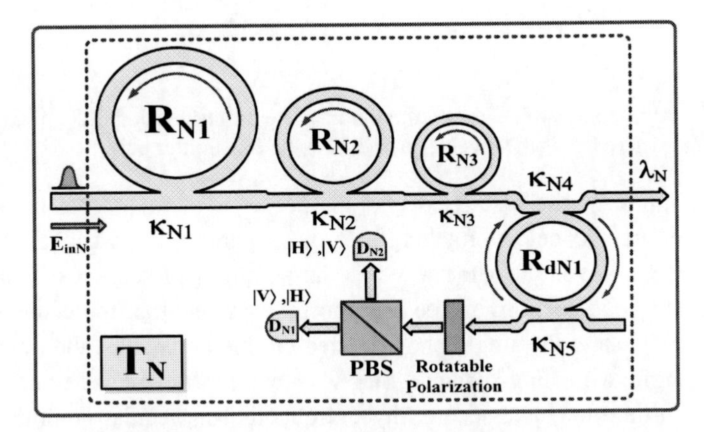

Figure 8.3. A schematic of a quantum tweezer generation system at the transmission unit (TN), where RNS is the ring radii, kNS is the coupling coefficients, RdNS is an add/drop ring radius, can be used to be the received part, PBS is the Polarizing Beamsplitter, DN is the Detectors.

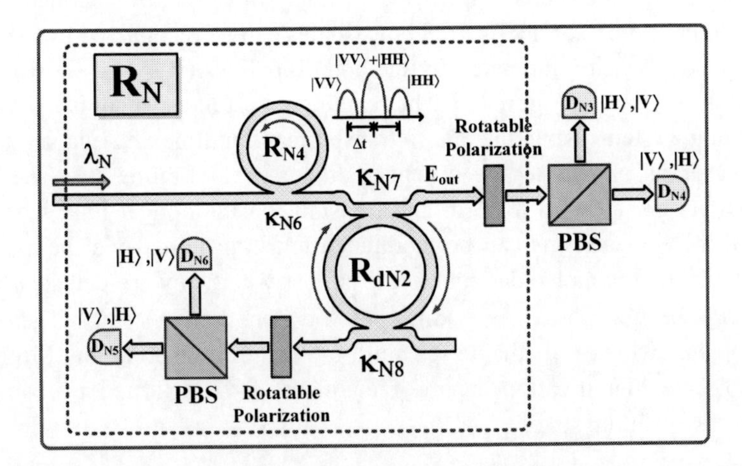

Figure 8.4. A schematic of an entangled photon pair manipulation within a ring resonator at the receiver unit (R_N). The quantum state is propagating to a rotatable polarizer and then is split by a beam splitter (PBS) flying to detector D_{N3} and D_{N4}.

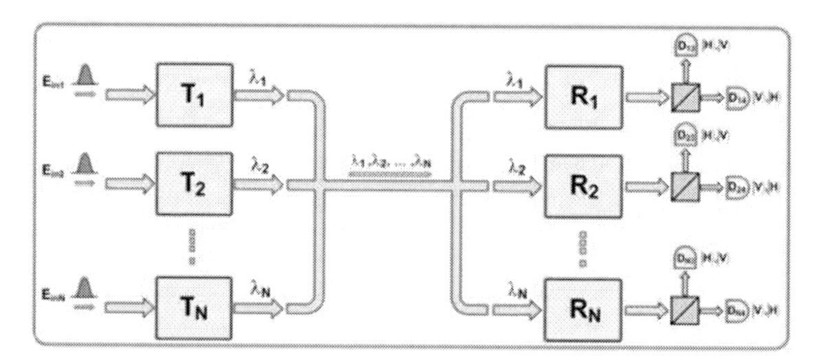

Figure 8.5. A schematic of a system of the transporters transmission system via an optical link, where RN is the receiver part, TN is the transmitter part.

For simplicity, the entangled photons power is attenuated to be a single photon be fore the detection, therefore, the separation between photon and molecule is employed the same way of a single photon detection scheme. This means that the detection of the transported single atom/molecule can be configured by using the single photon detection method. Thus, the transported molecule/atom with long distance link via quantum-molecular transporter is realized. Furthermore, the secured hybrid quantum-molecular communication can be implemented within the existed transmission link.

The transmitter unit (as shown in Figure 8.3) can be used to generate the quantum codes within the series of micro ring resonators and the cloning unit, which is operated by the add/drop filter (R_{dN1}). The receiver unit can be used to detect the quantum bits via the optical link, which can be obtained via the end quantum processor and the reference states can be recognized by using the cloning unit, which is operated by the add/drop filter (R_{dN2}) as shown in the schematic diagram in Figure 8.4. The remaining part of a system is the parallel processing system using the multi transporters multiplexing via an optical multiplexer as shown in the schematic diagram in Figure 8.5. The multi transporters are allowed to form and transmit via the optical link, where the trapped molecules/atoms can be modulated and transmitted via the available link, however, the molecular states will be lost when they are detected by the receiver. The transporters with different wavelengths (λ_N) can be generated and obtained which is available for multi molecules transportation. Moreover, the molecule identification can be recognized and confirmed by using the transporter quantum state.

8.5. CONCLUSION

We have demonstrated that some interesting results can be obtained when the laser pulse is propagated within the nonlinear optical ring resonator, especially, in microring and nanoring resonators, which can be used to perform many applications. For instance, the broad spectrum of a monochromatic source with reasonable power can be generated. By using a dark soliton, it can be converted to be a bright soliton by using the ring resonator system incorporating the add/drop multiplexer, which can be configured as a dynamic optical tweezers. The use of quantum tweezers for quantum-molecular communication/cryptography via a single photon based technology in the communication link is plausible.

REFERENCES

[1] K. Svoboda and S.M. Block, "Biological applications of optical forces, *"Annu. Rev. Biophys. Biomol. Struct*., 23, 247-285(1994).

[2] S.M. Block, "Making light work with optical tweezers." *Nature,* 360, 493-495(1992).

[3] R.M. Simmons, J.T. Finer, S. Chu, J.A. Spudich, "Quantitative measurements of force and displacement using an optical trap." *Biophys. J.,* 70(4), 1813-1822(1996).

[4] M.J. Lang, C.L. Asbury, J.W. Shaevitz, S.M. Block, "An automated two-dimensional optical force clamp for single molecule studies," *Biophys. J.,* 83(1), 491-501(2002).

[5] J. Pine, G. Chow, "Moving live dissociated neurons with an optical tweezer," *IEEE Transaction on Biomedical Engineering*, 56(4), 1184 – 1188(2009).

[6] A. Ashkin, J. M. Dziedzic and T. Yamane, "Optical trapping and manipulation of single cells using infrared laser beams", *Nature*, 330, 769-771(1987).

[7] J. Conia, B. S. Edwards and S. Voelkel, "The micro-robotic laboratory: optical trapping and scissing for the biologist," *J. Clin. Lab. Anal*., 11, 28–38(1997).

[8] N. Pornsuwancharoen, U. Dunmeekaew and P.P. Yupapin, "Multi-soliton generation using a microring resonator system for DWDM based

soliton communication," *Microw. and Opt. Technol. Lett.*, 51(5), 1374-1377(2009).

[9] P.P. Yupapin, N. Pornsuwanchroen and S. Chaiyasoonthorn, "Attosecond pulse generation using nonlinear microring resonators," *Microw. and Opt. Technol. Lett.,* 50(12), 3108-3011(2008).

[10] N. Pornsuwancharoen and P.P. Yupapin, "Generalized fast, slow, stop, and store light optically within a nanoring resonator," Microw. *and Opt. Technol. Lett.*, 51(4), 899-902(2009).

[11] N. Pornsuwancharoen, S. Chaiyasoonthorn and P.P. Yupapin, "Fast and slow lights generation using chaotic signals in the nonlinear microring resonators for communication security," *Opt. Eng.*, 48(1), 50005-1-5(2009).

[12] P.P. Yupapin and N. Pornsuwancharoen, "Proposed nonlinear microring resonator arrangement for stopping and storing light," *IEEE Photon. Technol. Lett.,* 21, 404-406(2009).

[13] K. Sarapat, N. Sangwara, K. Srinuanjan, P.P. Yupapin and N. Pornsuwancharoen, "Novel dark-bright optical solitons conversion system and power amplification," *Opt. Eng.,* 48, 045004-1-7(2009).

[14] S. Mitatha, N. Pornsuwancharoen and P.P.Yupapin, "A simultaneous short-wave and millimeter-wave generation using a soliton pulse within a nano-waveguide", *IEEE Photon. Technol. Lett.*, 21, 932-934 (2009).

[15] G. P. Agrawal, *Nonlinear Fiber Optics*, Academic Press, 4th edition, New York, 2007.

[16] M.E. Heidari, M.K. Moravvej-Farshi, and A. Zariffkar, "Multichannel wavelength conversion using fourth-order soliton decay", *J. Lightwave Technol.,* 25, 2571-2578 (2007).

[17] A. Charoenmee, N. Pornsuwancharoen and P.P.Yupapin, "Trapping a dark soliton pulse within a nanoring resonator", *International J. of Light and Electron Optics,* (2009). doi:10.1016/j.ijleo.2009.03.015.

[18] Y. Kokubun, Y. Hatakeyama, M. Ogata, S. Suzuki, and N. Zaizen, "Fabrication technologies for vertically coupled microring resonator with multilevel crossing busline and ultracompact-ring radius," *IEEE J. Sel. Top. Quantum Electron.* 11, 4–10(2005).

[19] Y. Su, F. Liu, and Q. Li, "System performance of slow-light buffering, and storage in silicon nano-waveguide," *Proc. SPIE* 6783, 67832P(2007).

[20] P.P. Yupapin, P. Saeung and C. Li, "Characteristics of complementary ring-resonator add/drop filters modeling by using graphical approach," *Opt. Commun.*, 272, 81-86(2007).

[21] P.P. Yupapin and W. Suwancharoen, "Chaotic signal generation and cancellation using a microring resonator incorporating an optical add/drop multiplexer," *Opt. Commun.*, 280/2, 343-350(2007).

[22] L. Yuan, Z. Liu, J. Yang and C. Guan, "Twin-core fiber optical tweezers", *Opt. Exp.*, 16, 4559-4566 (2008).

[23] N. Malagninoa, G. Pescea, A. Sassoa and E. Arimondo, "Measurements of trapping efficiency and stiffness in optical tweezers", *Opt. Commun.*, 214, 15-24 (2002).

[24] S. Suchat, W. Khannam and P.P. Yupapin, "Quantum key distribution via an optical wireless communication link for telephone network," *Opt. Eng. Lett.*, 46(10), 100502-1-5(2007).

[25] P.P. Yupapin and S. Suchat, "Entangle photon generation using fiber optic Mach-Zehnder interferometer incorporating nonlinear effect in a fiber ring resonator, *Nanophotonics (JNP*), 1, 13504-1-7(2007).

Chapter 9

NANOPHOTONIC CIRCUITS

9.1. INTRODUCTION

To date, many researchers have demonstrated the interesting techniques that can be used to realize the various optical logic functions (i.e. AND, NAND, OR, XOR, XNOR, NOR) by using different schemes, including thermo-optic effect in two cascaded microring resonator [1], quantum dot [2, 3], semiconductor optical amplifier (SOA) [4-11], TOAD-based interferometer device [12], nonlinear effects in SOI waveguide [13, 14], nonlinear loop mirror [15,16], DPSK format [17,18], local nonlinear in MZI [19], photonic crystal [20, 21], error correction in multipath differential demodulation [22], fiber optical parametric amplifier [23], multimode interference in SiGe/Si [24], polarization and optical processor [25], and injection-locking effect in semiconductor laser [26]. However, the searching of new techniques remain, there are some rooms for new techniques that can be used to be the good candidate. Therefore, in this chapter, we propose the use of the simultaneous arbitrary two-input logic XOR/XNOR and all-optical logic gates based on dark-bright soliton conversion within the add/drop optical filter system. The advantage of the scheme is that the random codes can be generated simultaneously by using the dark-bright soliton conversion behavior, in which the coincidence dark and bright soliton can be separated after propagating into a $\pi/2$ phase retarder, which can be used to form the security codes. Moreover, this is a simple and flexible scheme for an arbitrary logic switching system, which can be used to form the advanced complex logic circuits. The proposed scheme is based on a 1 bit binary comparison XOR/XNOR scheme that can be compared to any 2 bits, i.e., between 0 and 0 (dark-dark solitons), 0 and 1 (dark-bright solitons), 1 and 0 (bright-dark solitons) or 1 and 1 (bright-bright solitons), which will be detailed in the next section.

9.2. DARK-BRIGHT SOLITON CONVERSION MECHANISM

In operation, dark-bright soliton conversion using a ring resonator optical channel dropping filter (OCDF) is composed of two sets of coupled waveguides, as shown in Figure 9.1(a) and 9.1(b), where for convenience, Figure 9.1(b) is replaced by Figure 9.1(a). The relative phase of the two output light signals after coupling into the optical coupler is $\pi/2$ before coupling into the ring and the input bus, respectively. This means that the signals coupled into the drop and through ports are acquired a phase of π with respect to the input port signal. In application, if we engineer the coupling coefficients appropriately, the field coupled into the through port on resonance would completely extinguish the resonant wavelength, and all power would be coupled into the drop port. We will show that this is possible later in this section.

$$E_{ra} = -j\kappa_1 E_i + \tau_1 E_{rd}, \tag{9.1}$$

$$E_{rb} = \exp(j\omega T/2)\exp(-\alpha L/4) E_{ra}, \tag{9.2}$$

$$E_{rc} = \tau_2 E_{rb} - j\kappa_2 E_a, \tag{9.3}$$

$$E_{rd} = \exp(j\omega T/2)\exp(-\alpha L/4) E_{rc}, \tag{9.4}$$

$$E_t = \tau_1 E_i - j\kappa_1 E_{rd}, \tag{9.5}$$

$$E_d = \tau_2 E_a - j\kappa_2 E_{rb}, \tag{9.6}$$

where E_i is the input field, E_a is the add(control) field, E_t is the through field, E_d is the drop field, $E_{ra}...E_{rd}$ are the fields in the ring at points $a...d$, κ_1 is the field coupling coefficient between the input bus and ring, κ_2 is the field coupling coefficient between the ring and output bus, L is the circumference of the ring, T is the time taken for one round trip(roundtrip time), and α is the power loss in the ring per unit length. We assume that this is the lossless coupling, i.e., $\tau_{1,2} = \sqrt{1-\kappa_{1,2}^2}$. $T = L n_{eff}/c$.

The output power/intensities at the drop and through ports are given by

$$\left|E_d\right|^2 = \left|\frac{-\kappa_1\kappa_2 A_{1/2}\Phi_{1/2}}{1-\tau_1\tau_2 A\Phi}E_i + \frac{\tau_2 - \tau_1 A\Phi}{1-\tau_1\tau_2 A\Phi}E_a\right|^2 . \tag{9.7}$$

$$\left|E_t\right|^2 = \left|\frac{\tau_2 - \tau_1 A\Phi}{1-\tau_1\tau_2 A\Phi}E_i + \frac{-\kappa_1\kappa_2 A_{1/2}\Phi_{1/2}}{1-\tau_1\tau_2 A\Phi}E_a\right|^2 . \tag{9.8}$$

where $A_{1/2} = \exp(-\alpha L/4)$ (the half-round-trip amplitude), $A = A_{1/2}^2$, $\Phi_{1/2} = \exp(j\omega T/2)$ (the half-round-trip phase contribution), and $\Phi = \Phi_{1/2}^2$.

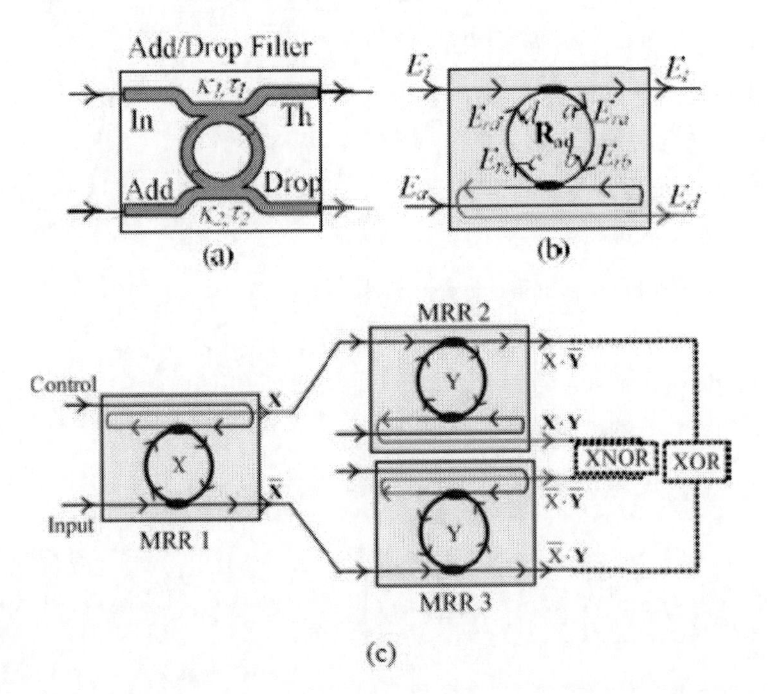

Figure 9.1. A schematic diagram of a simultaneous optical logic XOR and XNOR gate.

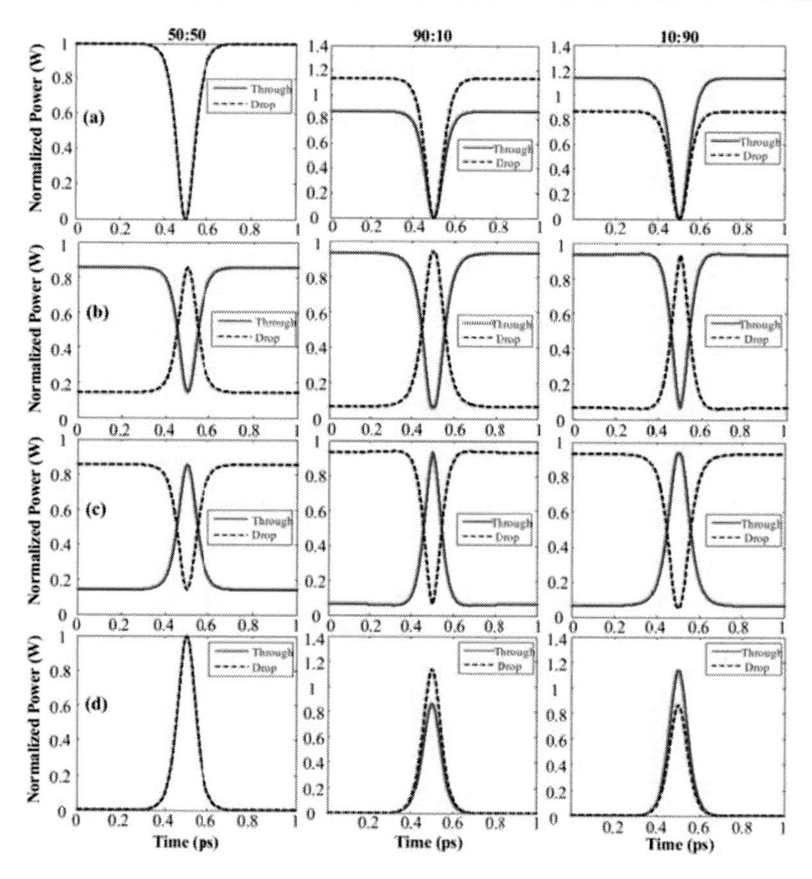

Figure 9.2. Dark-bright soliton conversion results.

The input and control fields at the input and add ports are formed by the dark-bright optical soliton as shown in Eqs. (9.9) – (9.10),

$$E_{in}(t) = A_0 \tanh\left[\frac{T}{T_0}\right]\exp\left[\left(\frac{z}{2L_D}\right) - i\omega_0 t\right] \tag{9.9}$$

$$E_{in}(t) = A_0 \operatorname{sech}\left[\frac{T}{T_0}\right]\exp\left[\left(\frac{z}{2L_D}\right) - i\omega_0 t\right] \tag{9.10}$$

where A_0 and z are the optical field amplitude and propagation distance, respectively. $T=t-\beta_1 z$, where β_1 and β_2 are the coefficients of the linear and second-order terms of Taylor expansion of the propagation constant. $L_D=$

$T_0^2/|\beta_2|$ is the dispersion length of the soliton pulse. T_0 in equation is a soliton pulse propagation time at initial input (or soliton pulse width), where t is the soliton phase shift time, and the frequency shift of the soliton is ω_0. When the optical field is entered into the nanoring resonator as shown in Figure 9.2, where the coupling coefficient ratio $\kappa_1:\kappa_2$ are *50:50, 90:10, 10: 90*. By using (a) dark soliton is input into input and control ports, (b) dark and bright soliton are used for input and control signals, (c) bright and dark soliton are used for input and control signals, and (d) bright soliton is used for input and control signals. The ring radii $R_{ad} = 300nm$, $A_{eff} = 0.25\mu m^2$, $n_{eff} = 3.14$(for InGaAsP/InP), $\alpha = 0.1dB/mm$, $\gamma = 0.01$, $\lambda_0 = 1.55\ \mu m$.

9.3. Optical XOR/XNOR Logic Gate Operation

The proposed architecture is schematically shown in Figure 9.1(c). A continuous optical wave with a wavelength of λ is formed by an optical dark-bright soliton pulse train X using MRR 1, in which the optical pulse trains that appear at the through and drop ports of MRR 1 are $\overline{X}$ and X, respectively ($\overline{X}$ is the inverse of X or dark-bright conversion). It is assumed that the input optical dark-bright soliton wave is directed to the drop port when the optical signal is 1(dark soliton pulse). In other words, the MRR 1 resonates at λ when the input dark soliton pulse is applied.

If the optical pulse train X is fed into MRR 2 from its input port solely and is formed by an optical pulse train Y bit by bit using MRR 2, where here, we assume that no signal is fed into MRR 2 from its add port, in which the optical pulse trains that appear at the through and drop ports of MRR 2 will be $X \cdot \overline{Y}$ and $X \cdot Y$, respectively, whereas the aforementioned assumption is provided. The symbol represents the logical operation AND here.

If the optical pulse train $\overline{X}$ is fed into MRR 3 from its input port solely and is formed by an optical pulse train Y bit by bit using MRR 3, where here, we assume that no signal is fed into MRR 3 from its input port, in which the optical pulse trains that appear at the through and drop ports of MRR 3 will be $\overline{X} \cdot Y$ and $\overline{X} \cdot \overline{Y}$, respectively. If the optical pulse trains X and $\overline{X}$ are fed into MRR 2 and MRR 3 from its input ports simultaneously [see Figure 9.1(c)], in which the optical pulse trains $X \cdot \overline{Y} + \overline{X} \cdot Y$ and $X \cdot Y + \overline{X} \cdot \overline{Y}$ are achieved at the through and drop ports of MRR 2 and MRR 3, respectively [see Figure 9.1(c)]. The symbol + represents the logical operation OR, which is implemented through the multiplexing function of MRR 2 and MRR 3. It is well known that the XOR and XNOR operations can be calculated by using the formulas $X \oplus Y = X \cdot \overline{Y} + \overline{X} \cdot Y$ and $\mathrm{X} \otimes \mathrm{Y} = \mathrm{X} \cdot \mathrm{Y} + \overline{\mathrm{X}} \cdot \overline{\mathrm{Y}}$, where the capital letters represent logical variables and the symbols ⊕ and ⊗ represent the XOR and XNOR operators, respectively. Therefore, the proposed architecture can be used as an XOR and XNOR calculator.

By using the dynamic performance of the device is shown in Figure 9.3, two pseudo-random binary sequence 2^4-1 signals at 100 Gbit/s are converted to be two optical signals bit by bit according to the rule presented above and then applied to the corresponding MRRs. Clearly, a logic 1 is obtained when the applied optical bright soliton pulse signals, and a logic 0 is obtained when the applied optical dark soliton pulse signals are generated. Therefore, the device performs the XOR and XNOR operation correctly.

The proposed simultaneous all-optical logic XOR and XNOR gates device is as shown in Figure 9.1(c). The input and control light pulse trains are input into the first add/drop optical filter (MRR 1) using the dark solitons (logic '0') or the bright solitons (logic '1'). Firstly, the dark soliton is converted to be dark and bright soliton via the add/drop optical filter, which they can be seen at the through and drop ports with π phase shift [27], respectively. By using the add/drop optical filters (MRR 2 and MRR 3), both input signals are generated by the first stage add/drop optical filter. Next, the input data "Y" with logic "0"(dark soliton) and logic "1"(bright soliton) are added into both add ports, the dark-bright soliton conversion with π phase shift is operated again. For large scale (Figure 9.1(c)), results obtained are simultaneously seen by D_1, D_2, T_1, and T_2 at the drop and through ports for optical logic XNOR and XOR gates, respectively.

In simulation, the add/drop optical filter parameters are fixed for all coupling coefficients to be $\kappa_s = 0.05$, $R_{ad} = 300nm$, $A_{eff} = 0.25\ \mu m^2$ [28], $\alpha = 0.05dBmm^{-1}$ for all add/drop optical filters in the system. Results of the simultaneous optical logic XOR and XNOR gates are generated by using dark-bright soliton conversion with wavelength center at $\lambda_0 = 1.50\mu m$, pulse width *35 fs*. In Figure 9.4, simulation result of the simultaneous output optical logic gate is seen when the input data logic "00" is added, whereas the obtained output optical logic is "0001" [see Figure 9.4(a)]. Similarly, when the simultaneous output optical logic gate input data logic "01" is added, the output optical logic "0010" is formed [see Figure 9.4(b)]. Next, when the output optical logic gate input data logic "10" is added, the output optical logic "1000" is formed [see Figure 9.4(c)]. Finally, when the output optical logic input data logic “11” is added, we found that the output optical logic “0100” is obtained [see Figure 9.4(d)].

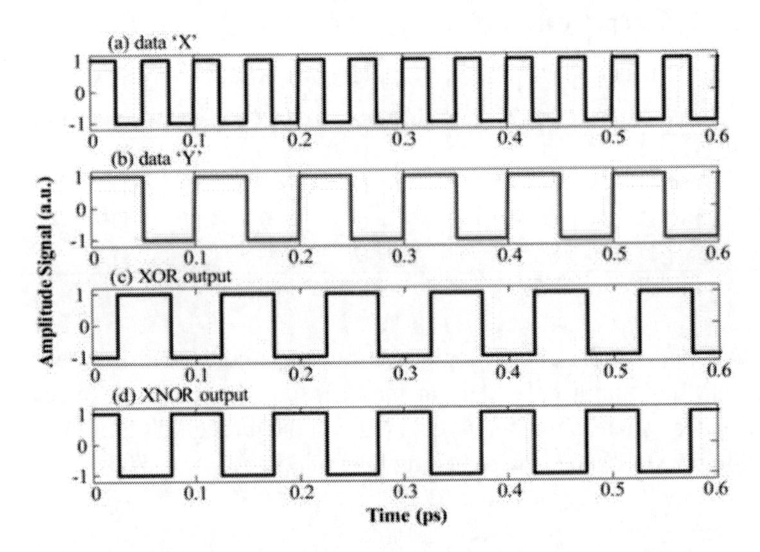

Figure 9.3. Output results dynamic performance of the device, when (a) data ‘X’, (b) data ‘Y’, (c) all-optical XOR and (d) XNOR logic gates.

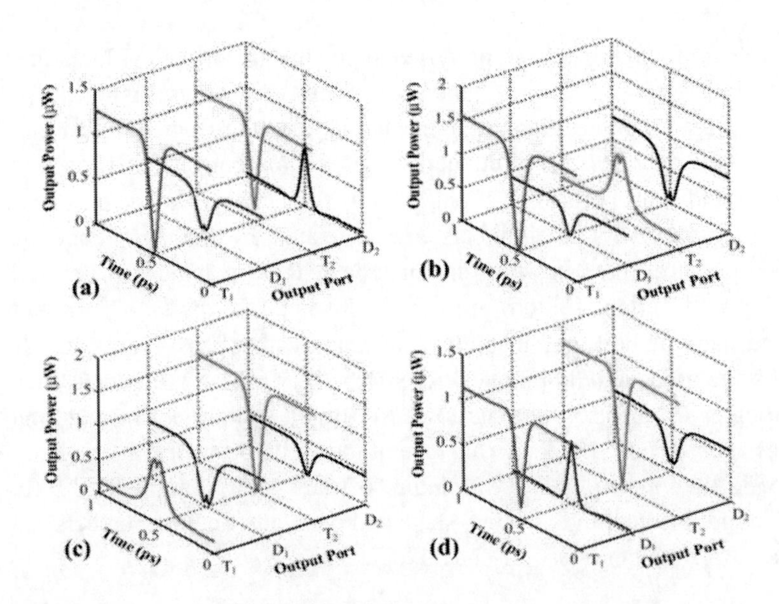

Figure 9.4. Shows the output logic XOR/XNOR gates when the input logic states are (a) ‘DD’, (b) ‘DB’, (c) ‘BD’, and (d) ‘BB’, respectively.

Table 9.1. Conclusion output optical logic XOR and XNOR gates

Input data		Output logic					
X	Y	(T_1) $X \cdot \overline{Y}$	(D_1) $X \cdot Y$	(T_2) $\overline{X} \cdot Y$	(D_2) $\overline{X} \cdot \overline{Y}$	XOR $X \cdot \overline{Y} + \overline{X} \cdot Y$	XNOR $X \cdot Y + \overline{X} \cdot \overline{Y}$
D	D	D	D	D	B	D	B
D	B	D	D	B	D	B	D
B	D	B	D	D	D	B	D
B	B	D	B	D	D	D	B

'D (Dark soliton)' = logic '0', 'B (Bright soliton)' = logic '1'.

The simultaneous optical logic gate output is concluded in Table 1. We found that the output data logic in the drop ports, D_1, D_2 are optical logic XNOR gates, whereas the output data logic in the through ports, T_1 and T_2 are optical logic XOR gates, the switching time of *35.1 fs* is noted.

9.4. Operation Principle of Simultaneous All-Optical Logic Gates

The configuration of the proposed simultaneous all-optical logic gates is shown in Figure 9.5. The input and control light ("A") pulse trains in the first add/drop optical filter (No. "01") are the dark soliton (logic '0'). In the first stage of the add/drop filter, the dark-bright soliton conversion is seen at the through and drop ports with π phase shift, respectively. In the second stage (No."11" and "12"), both inputs are generated by the first stage of the add/drop optical filter, in which the input data "B" with logic "0"(dark soliton) and logic "1"(bright soliton) are added into both add ports. The outputs of second stage are dark-bright soliton conversion with π phase shift again. In the third stage of the add/drop optical filter (No. "21" to "24"), the input data "C" with logic "0"(dark soliton) and logic "1"(bright soliton) are inserted into all final stage add ports. In the final stage of the add/drop optical filter (No. "31" to "38"), the input data "D" with logic "0"(dark soliton) and logic "1"(bright soliton) are inserted into all final stage add ports and outputs numbers 1 to 16 and shown simultaneously all-optical logic gate.

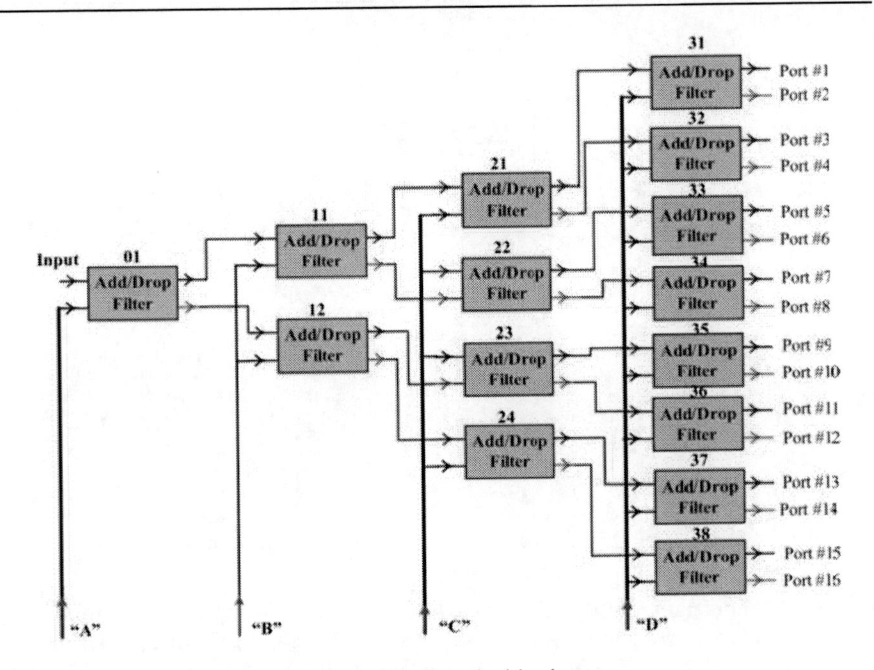

Figure 9.5. A schematic of the proposed all-optical logic gate.

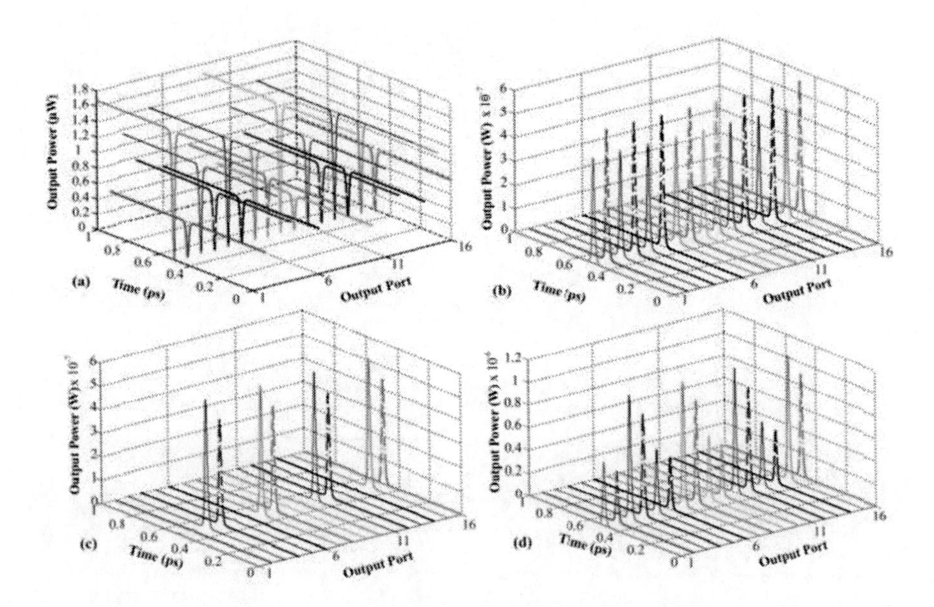

Figure 9.6 (Continues).

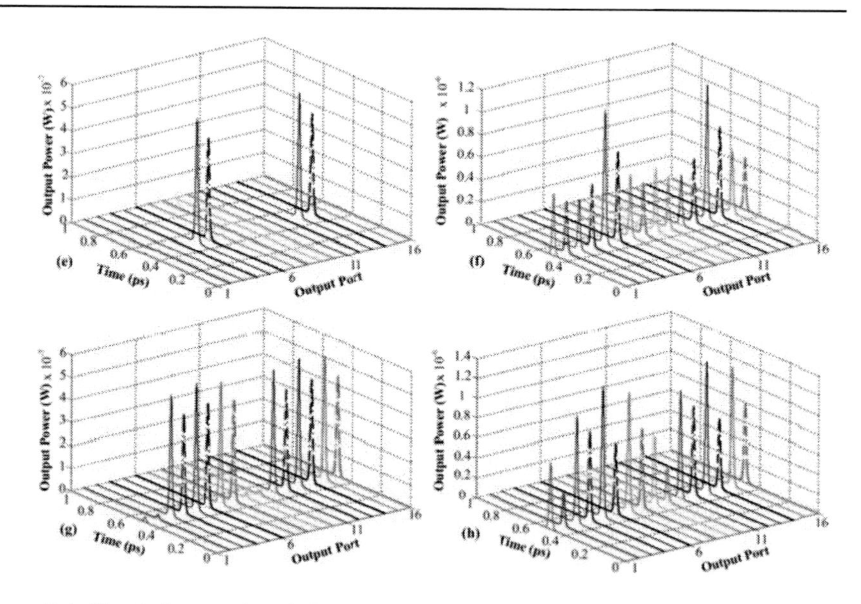

Figure 9.6. Simulation results of all-optical logic gate when the input and control signals ('*A*') is dark soliton. Input data '*BCD*' are (a) 'DDD', (b) 'DDB', (c) 'DBD', (d) 'DBB', (e) 'BDD', (f) 'BDB', (g) 'BBD' and (h) 'BBB', respectively.

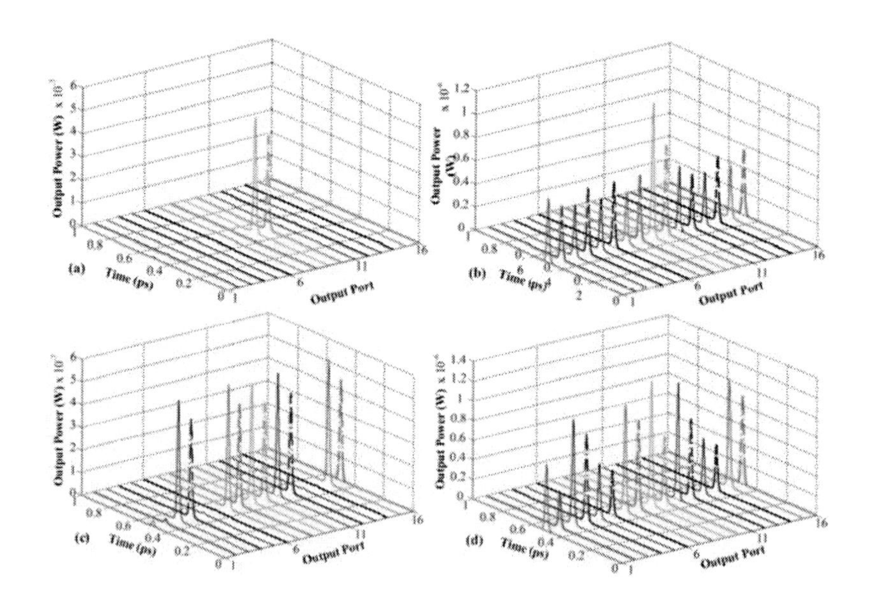

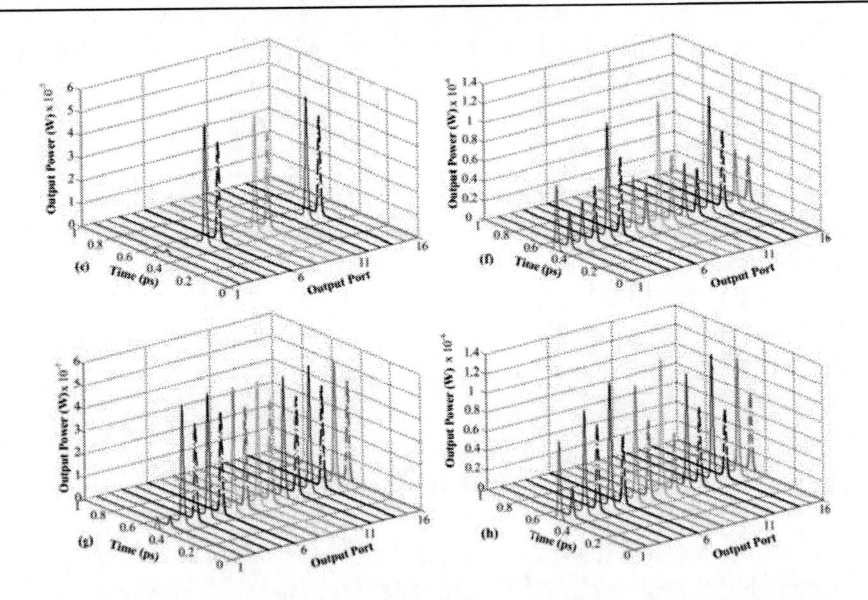

Figure 9.7. Simulation results of all-optical logic gate when the dark soliton input and control signal (‘*A*’) is bright soliton. The inputs ‘*BCD*’ are (a) ‘DDD’, (b) ‘DDB’, (c) ‘DBD’, (d) ‘DBB’, (e) ‘BDD’, (f) ‘BDB’, (g) ‘BBD’ and (h) ‘BBB’, respectively.

Table 9.2. Conclusion output results of all-optical logic gates

“A”	Output Port No.															
	1	2	3	4	5	6	7	8	9	10	11	12	13	14	15	16
DARK (‘0’)	0	0	0	0	0	0	0	0	0	0	0	0	0	0	0	0
	0	1	0	1	0	1	0	1	0	1	0	1	0	1	0	1
	0	0	1	0	0	0	1	0	0	0	1	0	0	0	1	0
	1	0	1	1	1	0	1	1	1	0	1	1	1	0	1	1
	0	0	0	0	1	0	0	0	0	0	0	0	1	0	0	0
	1	0	0	1	1	1	1	0	1	0	0	1	1	1	1	0
	0	0	1	0	1	0	1	0	0	0	1	0	1	0	1	0
	1	0	1	1	1	1	1	1	1	0	1	1	1	1	1	1
BRIGHT (‘1’)	0	0	0	0	0	0	0	0	1	0	0	0	0	0	0	0
	1	0	0	1	0	1	0	1	1	1	1	0	0	1	0	1
	0	0	1	0	0	0	1	0	1	0	1	0	0	0	1	0
	1	0	1	1	1	0	1	1	1	1	1	1	1	0	1	1
	0	0	0	0	1	0	0	0	1	0	0	0	1	0	0	0
	1	0	0	1	1	1	1	0	1	1	1	0	1	1	1	0
	0	0	1	0	1	0	1	0	1	0	1	0	1	0	1	0
	1	0	1	1	1	1	1	1	1	1	1	1	1	1	1	1

AND, NOR, XNOR, XOR, NAND, OR

The simulation parameters of the add/drop optical filters are fixed for all coupling coefficients to be $\kappa_s = 0.05$, $R_{ad} = 300nm$, $A_{eff} = 0.25\ \mu m^2$ [27], and $\alpha = 0.05dBmm^{-1}$ for all add/drop optical filter system. Simulation results of the simultaneous all-optical logic gates are generated by using the dark-bright soliton conversion at wavelength center $\lambda_0 = 1.50\mu m$, pulse width *35 fs*. In Figure 9.6, the simulation result of simultaneous output optical logic gate when the input and control signals (‘*A*’) are dark solitons. Input data ‘*BCD*’ are (a) ‘DDD’, (b) ‘DDB’, (c) ‘DBD’, (d) ‘DBB’, (e) ‘BDD’, (f) ‘BDB’, (g) ‘BBD’ and (h) ‘BBB’, respectively. Results of all outputs are concluded as shown in table 2 for all-optical logic gates.

9.5. Conclusion

We have proposed the novel technique that can be used to simultaneously generate the optical logic AND and OR gates using dark-bright soliton conversion within the add/drop optical filter system. By using the dark-bright soliton conversion concept, results obtained have shown that the input logic ‘0’ and control logic ‘0’ can be formed by using the dark soliton (D) pulse trains. We also found that the simultaneous optical logic AND and OR gates can be seen randomly at the drop and through ports, respectively, which is shown the potential of application for large scale use, especially, for security code application requirement.

References

[1] L. Zhang, R. Ji, L. Jia, L. Yang, P. Zhou, Y. Tian, P. Chen, Y. Lu, Z. Jiang, Y. Liu, Q. Fang, and M. Yu, “Demonstration of directed XOR/XNOR logic gates using two cascaded microring resonators,” *Opt. Lett.*, 35(10), 1620-1622(2010).

[2] S. Ma, Z. Chen, H. Sun, and K. Dutta, “High speed all optical logic gates based on quantum dot semiconductor optical amplifiers,” *Opt. Express,* 18(7), 6417-6422(2010).

[3] T. Kawazoe, K. Kobayashi, K. Akahane, M. Naruse, N. Yamamoto, and M. Ohtsu, “Demonstration of nanophotonic NOT gate using near-field optically coupled quantum dots,” *Appl. Phys. B*, 84, 243-246(2006).

[4] J. Dong, X. Zhang, and D. Huang, "A proposal for two-input arbitrary Boolean logic gates using single semiconductor optical amplifier by picoseconds pulse injection," *Opt. Express*, 17(10), 7725-7730(2009).
[5] J. Dong, X. Zhang, J. Xu, and D. Huang, "40 Gb/s all-optical logic NOR and OR gates using a semiconductor optical amplifier: Experimental demonstration and theoretical analysis," *Opt. Commun.* 281, 1710-1715 (2008).
[6] B. C. Han, J. L. Yu, W. R. Wang, L. T. Zhang, H. Hu, and E. Z. Yang, "Experimental study on all-optical half-adder based on semiconductor optical amplifier," *Optoelectron. Lett.*, 5(3), 0162-0164(2009).
[7] Z. Li, Y. Liu, S. Zhang, H. Ju, H. de Waardt, G. D. Khoe, H. J. S. Dorren, and D. Lenstra, "All-optical logic gates using semiconductor optical amplifier assisted by optical filter," *Electron. Lett.* 41, 1397-1399 (2005).
[8] S. H. Kim, J. H. Kim, B. G. Yu, Y. T. Byun, Y. M. Jeon, S. Lee, and D. H. Woo, "All-optical NAND gate using cross-gain modulation in semiconductor optical amplifiers," *Electron. Lett.* 41, 1027-1028 (2005).
[9] X. Zhang, Y. Wang, J. Sun, D. Liu, and D. Huang, "All-optical AND gate at 10 Gbit/s based on cascaded single-port-couple SOAs," *Opt. Express* 12, 361-366 (2004).
[10] Z. Li, and G. Li, "Ultrahigh-speed reconfigurable logic gates based on four-wave mixing in a semiconductor optical amplifier," *IEEE Photon. Technol. Lett.* 18, 1341-1343 (2006).
[11] S. Kumar, and A. E. Willner, "Simultaneous four-wave mixing and cross-gain modulation for implementing an all-optical XNOR logic gate using a single SOA," *Opt. Express* 14, 5092-5097 (2006).
[12] J. N. Roy, and D. K. Gayen, "Integrated all-optical logic and arithmetic operations with the help of a TOAD-based interferometer device-alternative approach," *Appl. Opt.*, 46(22), 5304-5310(2007).
[13] M. Khorasaninejad, and S. S. Saini, "All-optical logic gates using nonlinear effects in silicon-on-insulator waveguides," *Appl. Opt.*, 48(25), F31-F36(2009).
[14] Y. D. Wu, "All-optical logic gates by using multibranch waveguide structure with localized optical nonlinearity," *IEEE J. Select. Quant. Electron.*, 11(2), 307-312(2005).
[15] Y. Miyoshi, K. Ikeda, H. Tobioka, T. Inoue, S. Namiki, and K. I. Kitayama, "Ultrafast all-optical logic gate using a nonlinear optical loop mirror based multi-periodic transfer function," *Opt. Express,* 16(4), 2570-2577(2008).

[16] T. Houbavlis, K. Zoiros, A. Hatziefremidis, H. Avramopoulos, L. Occhi, G. Guekos, S. Hansmann, H. Burkhard, and R. Dall'Ara, "10Gbit/s all-optical Boolean XOR with SOA fibre Sagnac gate," *Electron. Lett.* 35, 1650-1652 (1999).

[17] J. Wang, Q. Sun, and J. Sun, "All-optical 40 Gbit/s CSRZ-DPSK logic XOR gate and format conversion using four-wave mixing," *Opt. Express,* 17(15), 12555-12563(2009).

[18] J. Xu, X. Zhang, Y. Zhang, J. Dong, D. Liu, and D. Huang, "Reconfigurable all-optical logic gates for multi-input differential phase-shift keying signals: design and experiments," *J. Lightwave Technol.*, 27(23), 5268-5275(2009).

[19] Y. D. Wu and T. T. Shih, "New all-optical logic gates based on the local nonlinear Mach-Zehnder interferometer," *Opt. Express*, 16(1), 248-257(2008).

[20] Y. Zhang, Y. Zhang, and B. Li, "Optical switches and logic gates based on self-collimated beams in two-dimensional photonic crystals," *Opt. Express,* 15(15), 9287-9292(2007).

[21] G. Berrettini, A. Simi, A. Malacarne, A. Bogoni, and L. Poti, "Ultrafast integrable and reconfigurable XNOR, AND, NOR, and NOT photonic logic gate," *IEEE Photon. Technol. Lett.*, 18(8), 917-919(2006).

[22] Y. K. Lize, L. Christen, M. Nazarathy, S. Nuccio, X. Wu, A. E. Willner, and R. Kashyap, "Combination of optical and electronic logic gates for error correction in multipath differential demodulation," *Opt. Express,* 15(11), 6831-6839(2007).

[23] D. M. F. Lai, C. H. Kwok, and K. K. Y. Wong, "All-optical picoseconds logic gates based on a fiber optical parametric amplifier," *Opt. Express,* 16(22), 18362-18370(2008).

[24] Z. Li, Z. Chen, and B. Li, "Optical pulse controlled all-optical logic gates in SiGe/Si multimode interference," *Opt. Express*, 13(3), 1033-1038(2005).

[25] Y. A. Zaghloul and A. R. M. Zaghloul, "Complete all-optical processing polarization-based binary logic gates and optical processors," *Opt. Express,* 14(21), 9879-9895(2006).

[26] L. Y. Han, H. Y. Zhang, and Y. L. Guo, "All-optical NOR gate based on injection-locking effect in a semiconductor laser," *Optoelectron. Lett.,* 4(1), 0034-0037(2008).

[27] S. Mookherjea and M. A. Schneider, "The nonlinear microring add-drop filter," *Opt. Express* 16, 15130-15136 (2008).

[28] Y. Kokubun, Y. Hatakeyama, M. Ogata, S. Suzuki and N. Zaizen, "Fabrication technologies for vertically coupled microring resonator with multilevel crossing busline and ultracompact-ring radius," *IEEE J. Sel. Top. Quantum Electron.,* 11, 4-10(2005).

Chapter 10

DRUG DELIVERY AND DIAGNOSIS

10.1. INTRODUCTION

Much effort has been made to explain the nature of optical forces and to describe them quantitatively by the establishment of theoretical models. This requires a detailed knowledge of the incident electromagnetic field that impinges on the object and of the properties of the object, which changes the incident field distribution by scattering, absorption or reemission of photons [1].The theory of optical vortices soliton developed to date assumes a background of constant amplitude and is able to capture a number of interesting features of vortices dynamics. Understanding this dynamics is important for future application of vortices soliton in steerable all optical switching devices based on the concept of dark /bright soliton [2]. The promising techniques of microscopic volume trapping and transportation within the add/drop multiplexer have been reported in both theory [3] and experiment [4], respectively. To date, the optical tweezer technique has become a powerful tool for manipulation of micrometer-sized particles in three spatial dimensions. Initially, the useful static tweezers are recognized, and the dynamic tweezers are now realized in practical works [5-7]. Schulz et al [8] have shown that the transfer of trapped atoms between two optical potentials could be performed. In principle, an optical tweezers use forces exerted by intensity gradients in the strongly focused beams of light to trap and move the microscopic volumes of matters. Moreover, the other combination of force is induced by the interaction between photons, which is caused by the photon scattering effects. In application, the field intensity can be adjusted and tuned to form the suitable trapping potential, in which the desired gradient field and scattering force can be formed the suitable trapping force.

In biological applications of optical trapping and manipulation, it is possible to remotely apply controlled forces on living cells, internal parts of cells, and large biological molecules without inflicting detectable optical damage. This has resulted in many unique applications, where one of the most important of them is in the drug delivery and trapping study [9]. Several researchers have developed the drug delivery system, which can be designed to improve the pharmacological and therapeutic. There is attempt seeks the way cures that be able to achieve through specific surface receptors and fast, where the controlled drug delivery can help parenteral and route drug delivery. For the conventional methods of drug delivery management, usually, drug delivery at constant controlled rate is preferable. However, a better method may be in response to the physiological needs of the body. Moreover, the several advantages over conventional molecules are nanoparticle networks that have been proposed for controlling the release of high molecular weight biomolecules and drug molecules [10-15].

In this chapter, we propose the use of drug delivery system using the optical vortices within a PANDA ring resonator in which the dynamic optical vortices are generated using a dark soliton, bright soliton and Gaussian pulse propagating within an add/drop optical multiplexer incorporating two nanoring resonators (PANDA ring resonator). The dynamic behaviors of solitons and Gaussian pulses are analyzed and described. To increase the channel multiplexing, the dark solitons with slightly different wavelengths are controlled and amplified within the tiny system. The trapping force stability is simulated and seen when the Gaussian pulse is used to control via the add (control) port. In application, the optical vortices (dynamic tweezers) can be used to store (trap) photon, atom, molecule, DNA, ion, or particle, which can perform the dynamic tweezers. By using the hybrid transceiver, where the transmitter and receiver parts can be integrated by a single system. Here, the use of the transceiver to form the hybrid communication of those microscopic volumes of matters in the nanoscale regime can be realized, especially, for drug delivery application.

10.2. Optical Vortex Generation

In operation, the optical tweezers use force that are exerted by the intensity gradients in the strongly focused beam of light to trap and move the microscopic volume of matter, in which the optical force are customarily and may be described for a trapped mass/volume that give a viscous damping defined by the relationship equate optical force with Stokes force as follow [16, 17].

$$F = \frac{Qn_m P}{c} = \gamma_0 \dot{x} \quad (10.1)$$

where

$$\gamma_0 = 6\pi r \rho v \quad (10.2)$$

where $\dot{x}$ is velocities of volume, n_m is the index of refraction of the suspending medium, c is the speed of light, and P is the incident laser power, measured at the specimen. γ_0 is the stokes' drag term (viscous damping), r is a particle/volume radius, v is kinematic viscosity and ρ is fluids of density, Q is a dimensionless efficiency. Q represents the fraction of power utilized to exert force. For plane waves incident on a perfectly absorbing particle, Q = 1. To achieve stable trapping, the radiation pressure must create a stable, three-dimensional equilibrium. Because biological specimens are usually contained in aqueous medium, the dependence of F on nm can rarely be exploited to achieve higher trapping forces. Increasing the laser power is possible, but only over a limited range due to the possibility of optical damage. Q itself is therefore the main determinant of trapping force. It depends upon the NA, laser wavelength, light polarization state, laser mode structure, relative index of refraction, and geometry of the particle.

In the Rayleigh regime, trapping forces decompose naturally into two components. Since, in this limit, the electromagnetic field is uniform across the dielectric, particles can be treated as induced point dipoles. The scattering force is given by

$$F_{scatt} = \frac{\langle S \rangle \sigma}{c} \quad (10.3)$$

where

$$\sigma = \frac{8}{3}\pi(kr)^4 r^2 \left(\frac{m^2 - 1}{m^2 + 2}\right)^2 \qquad (10.4)$$

where is the scattering cross section of a Rayleigh sphere with radius r. S is the time-averaged Poynting vector, n is the index of refraction of the particle, $m = n / n_m$ is the relative index, and $k = 2\pi n_m / \lambda$ is the wave number of the light. Scattering force is proportional to the energy flux and points along the direction of propagation of the incident light. The gradient force is the Lorentz force acting on the dipole induced by the light field. It is given by

$$F_{grad} = \frac{\alpha}{2}\nabla\langle E^2\rangle \qquad (10.5)$$

where

$$\alpha = n_m^2 r^3 \left(\frac{m^2 - 1}{m^2 + 2}\right) \qquad (10.6)$$

is the polarizability of the particle. The gradient force is proportional and parallel to the gradient in energy density (for $m > 1$). The large gradient force is formed by the large depth of the laser beam, in which the stable trapping requires that the gradient force in the $-\hat{z}$ direction, which is against the direction of incident light (dark soliton valley), and it is greater than the scattering force. By increasing the NA, when the focal spot size is decreased, the gradient strength is increased [18], which can be formed within the tiny system, for instance, nanoscale device (nanoring resonator).

The trapping force is formed by using a dark soliton, in which the valley of the dark soliton is generated and controlled within the PANDA ring resonator by the control port signals. From Figure 10.1, the output field (E_{t1}) at the through port is given by [18]. We are looking for the system that can generate the dynamic tweezers (optical vortices), in which the microscopic volume can be trapped and transmission via the communication link. Firstly, the stationary and strong pulse that can propagate within the dielectric material (waveguide) for period of time is required. Moreover, the gradient field is an important property required in this case. Therefore, a dark soliton is satisfied and recommended to perform those requirements. Secondly, we are looking for the device that optical tweezers can propagate and form the long distance link, in which the gradient field (force) can be transmitted and received by using the same device. Here, the add/drop multiplexer in the form of a PANDA ring resonator which is well known and is introduced for this proposal, as shown in Figs. 10.1 and 10.2. To form the multi function operations, for instance, control, tune, amplify, the additional pulses are bright soliton and Gaussian pulse introduced into the system. The input optical field (E_{in}) and the add port optical field (E_{add}) of the dark soliton, bright soliton and Gaussian pulses are given by [19], respectively.

$$E_{in}(t) = A \tanh\left[\frac{T}{T_0}\right] \exp\left[\left(\frac{Z}{2L_D}\right) - i\omega_0 t\right] \tag{10.7}$$

$$E_{control}(t) = A \sec h\left[\frac{T}{T_0}\right] \exp\left[\left(\frac{Z}{2L_D}\right) - i\omega_0 t\right] \tag{10.8}$$

$$E_{control}(t) = E_0 \exp\left[\left(\frac{Z}{2L_D}\right) - i\omega_0 t\right] \tag{10.9}$$

where A and z are the optical field amplitude and propagation distance, respectively. T is a soliton pulse propagation time in a frame moving at the group velocity, $T=t-\beta_1 z$, where β_1 and β_2 are the coefficients of the linear and second-order terms of Taylor expansion of the propagation constant. $L_D= T_0^2/|\beta_2|$ is the dispersion length of the soliton pulse. T_0 in equation is a soliton pulse propagation time at initial input (or soliton pulse width), where t is the soliton phase shift time, and the frequency shift of the soliton is ω_0. This solution describes a pulse that keeps its temporal width invariance as it propagates, and thus is called a temporal soliton. When a soliton of peak intensity $\left(|\beta_2/\Gamma T_0^2|\right)$ is given, then T_0 is known. For the soliton pulse in the microring device, a balance should be achieved between the dispersion length (L_D) and the nonlinear length ($L_{NL}=1/\Gamma\phi_{NL}$). Here $\Gamma=n_2k_0$, is the length scale over which dispersive or nonlinear effects makes the beam become wider or narrower. For a soliton pulse, there is a balance between dispersion and nonlinear lengths. Hence $L_D = L_{NL}$. For a Gaussian pulse in Eq. (10.9), E_0 is the amplitude of optical field.

When light propagates within the nonlinear medium, the refractive index (n) of light within the medium is given by

$$n = n_0 + n_2 I = n_0 + \frac{n_2}{A_{eff}} P \tag{10.10}$$

with n_0 and n_2 as the linear and nonlinear refractive indexes, respectively. I and P are the optical intensity and the power, respectively. The effective mode core area of the device is given by A_{eff}. For the add/drop optical filter design, the effective mode core areas range from 0.10 to 0.50 μm^2, in which the parameters were obtained by using the related practical material parameters [(InGaAsP/InP) [20]. When a dark soliton pulse is input and propagated within a add/drop optical filter as shown in figure 1, the resonant output is formed. Thus, the normalized output of the light field is defined as the ratio between the output and input fields [$E_{\text{out}}(t)$ and $E_{\text{in}}(t)$] in each roundtrip. This is given as [21].

$$\left|\frac{E_{out}(t)}{E_{in}(t)}\right|^2 = (1-\gamma)\left[1 - \frac{(1-(1-\gamma)x^2)\kappa}{(1-x\sqrt{1-\gamma}\sqrt{1-\kappa})^2 + 4x\sqrt{1-\gamma}\sqrt{1-\kappa}\sin^2(\frac{\phi}{2})}\right] \tag{10.11}$$

The close form of Eq. (10.11) indicates that a ring resonator in this particular case is very similar to a Fabry–Perot cavity, which has an input and output mirror with a field reflectivity, $(1-\kappa)$, and a fully reflecting mirror. κ is the coupling coefficient, and $x=exp(-\alpha L/2)$ represents a roundtrip loss coefficient, $\phi_0=kLn_0$ and $\phi_{NL}=kLn_2|E_{in}|^2$ are the linear and nonlinear phase shifts, $k=2\pi/\lambda$ is the wave propagation number in a vacuum. L and α are the waveguide length and linear absorption coefficient, respectively. In this work, the iterative method is introduced to obtain the resonant results and similarly, when the output field is connected and input into the other ring resonators.

In order to retrieve the required signals, we propose to use the add/drop device with the appropriate parameters. This is given in the following details. The optical circuits of ring resonator add/drop filters for the through port and drop port can be given by Eqs. (10.12) and (10.13), respectively [22].

$$\left|\frac{E_t}{E_{in}}\right|^2=\frac{\left[(1-\kappa_1)+(1-\kappa_2)e^{-\alpha L}-2\sqrt{1-\kappa_1}\cdot\sqrt{1-\kappa_2}e^{-\frac{\alpha}{2}L}\cos(k_nL)\right]}{\left[1+(1-\kappa_1)(1-\kappa_2)e^{-\alpha L}-2\sqrt{1-\kappa_1}\cdot\sqrt{1-\kappa_2}e^{-\frac{\alpha}{2}L}\cos(k_nL)\right]} \quad (10.12)$$

$$\left|\frac{E_d}{E_{in}}\right|^2=\frac{\kappa_1\kappa_2e^{-\frac{\alpha}{2}L}}{1+(1-\kappa_1)(1-\kappa_2)e^{-\alpha L}-2\sqrt{1-\kappa_1}\cdot\sqrt{1-\kappa_2}e^{-\frac{\alpha}{2}L}\cos(k_nL)} \quad (10.13)$$

Here E_t and E_d represent the optical fields of the through port and drop ports, respectively. $\beta = kn_{\text{eff}}$ is the propagation constant, n_{eff} is the effective refractive index of the waveguide, and the circumference of the ring is $L=2\pi R$, with R as the radius of the ring. The filtering signal can be managed by using the specific parameters of the add/drop device, and the required signals can be retrieved via the drop port output. κ_1 and κ_2 are the coupling coefficients of the add/drop filters, $k_n=2\pi/\lambda$ is the wave propagation number for in a vacuum, and the waveguide (ring resonator) loss is $\alpha = 0.5$ dBmm^{-1}. The fractional coupler intensity loss is $\gamma = 0.1$. In the case of the add/drop device, the nonlinear refractive index is not effect to the system, therefore, it is neglected.

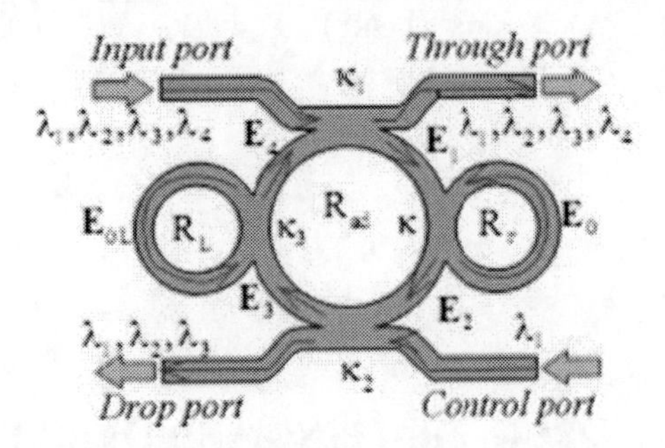

Figure 10.1. Schematic diagram of a proposed PANDA ring resonator.

From Eq. (10.12), the output field (E_{t1}) at the through port is given by

$$E_{t1}=AE_{i1}-BE_{i2}e^{-\frac{\alpha L}{2}-jk_n\frac{L}{2}}-\left[\frac{CE_{i1}e^{-\alpha L-jk_nL}+DE_{i2}e^{-\frac{3\alpha L}{2}-jk_n\frac{3L}{2}}}{1-Ee^{-\alpha L-jk_nL}}\right], \quad (10.14)$$

where

$A=\sqrt{(1-\gamma_1)(1-\gamma_2)}$, $B=\sqrt{(1-\gamma_1)(1-\gamma_2)\kappa_1(1-\kappa_2)}E_{0L}$, $C=\kappa_1(1-\gamma_1)\sqrt{(1-\gamma_2)\kappa_2}E_0E_{0L}$,

$D=(1-\gamma_1)(1-\gamma_2)\sqrt{\kappa_1(1-\kappa_1)\kappa_2(1-\kappa_2)}E_0E_{0L}^2$ and $E=\sqrt{(1-\gamma_1)(1-\gamma_2)(1-\kappa_1)(1-\kappa_2)}E_0E_{0L}$

The electric fields E_0 and E_{0L} are the field circulated within the nanoring at the right and left side of add/drop optical filter.

The power output (P_{t1}) at through port is written as

$$P_{t1}=|E_{t1}|^2. \tag{10.15}$$

The output field (E_{t2}) at drop port is expressed as

$$E_{t2}=\sqrt{(1-\gamma_2)(1-\kappa_2)}E_{i2}-\left[\frac{\sqrt{(1-\gamma_1)(1-\gamma_2)\kappa_1\kappa_2}E_0E_{i1}e^{-\frac{\alpha L}{2}-jk_n\frac{L}{2}}+XE_0E_{0L}E_{i2}e^{-\alpha L-jk_nL}}{1-YE_0E_{0L}e^{-\alpha L-jk_nL}}\right], \tag{10.16}$$

where

$$X=(1-\gamma_2)\sqrt{(1-\gamma_1)(1-\kappa_1)\kappa_2(1-\kappa_2)},\ Y=\sqrt{(1-\gamma_1)(1-\gamma_2)(1-\kappa_1)(1-\kappa_2)}$$

The power output (P_{t2}) at drop port is

$$P_{t2}=|E_{t2}|^2. \tag{10.17}$$

10.3. Drug Trapping and Delivery

By using the proposed design, the optical tweezers can be generated, trapped, transported and stored within the PANDA ring resonator and wavelength router as shown in Figs. 10.1 and 10.2, which can be used to form the microscopic volume transportation, i.e. drug delivery, via the waveguide [3, 4]. Simulation results of the dynamic optical vortices within the PANDA ring are as shown in Figure 10.3. In this case the bright soliton is input into the control port, and the trapped atoms/molecules are as shown in Figure 10.3(a)-10.3(f).

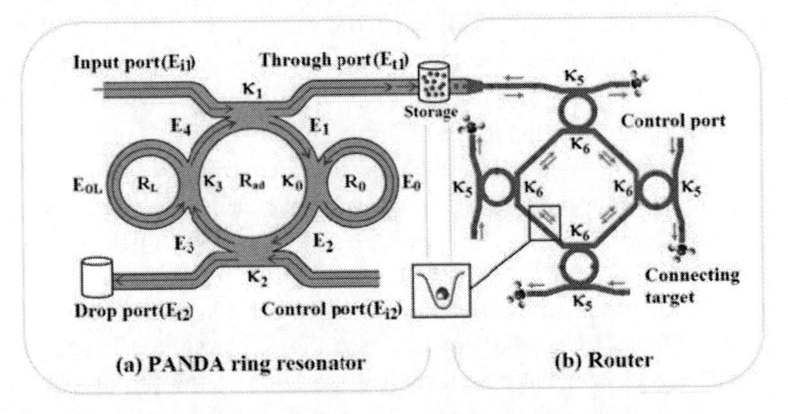

Figure 10.2. A system of drug delivery and distribution using optical vortices.

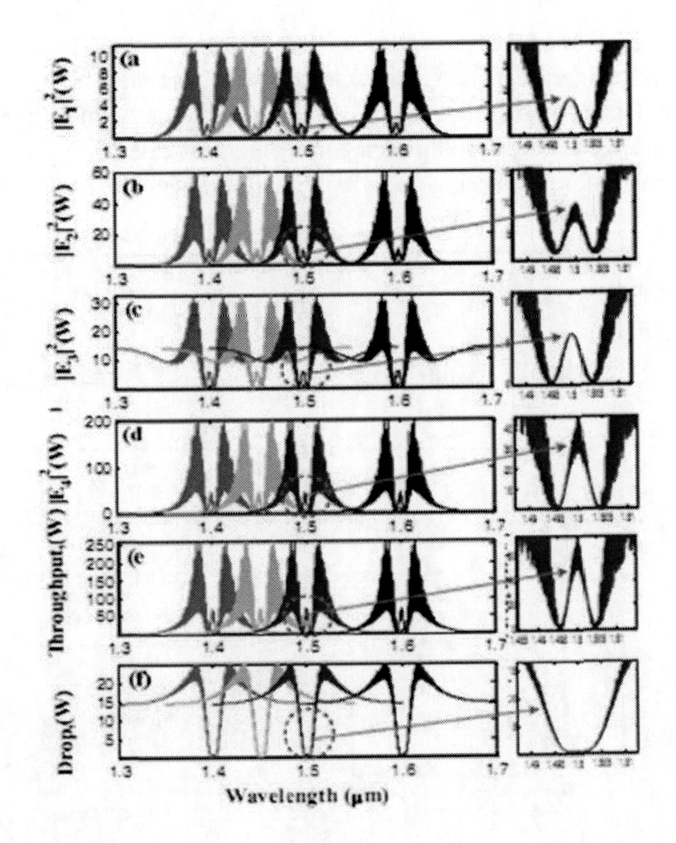

Figure 10.3. Simulation result of four potential wells (optical vortices/tweezers) with four different center wavelengths.

The ring radii are R_{add} = 1μm, R_R = 100 nm and R_L = 100 nm, in which the evidence of the practical device was reported by the authors in reference [23]. A_{eff} are 0.50, 0.25 and 0.25 μm^2: In this case, the dynamic tweezers (gradient fields) can be in the forms of bright soliton, Gaussian pulses and dark soliton, which can be used to trap the required microscopic volume. There are five different center wavelengths of tweezers generated, where the dynamical movements are (a) $|E_1|^2$, (b) $|E_2|^2$, (c) $|E_3|^2$, (d) $|E_4|^2$, (e) through port and (f) drop port signals.

More results of the optical tweezers generated within the PANDA ring are as shown in Figure 10.4, where in this case the bright soliton is used as the control port signal, and the output optical tweezers of the through and drop ports with different coupling constants are as shown in Figure 10.4(e) and 10.4(f), respectively, the coupling coefficients are (1) 0.1, (2) 0.35, (3) 0.6 and (4) 0.75.

The important aspect of the result is that the tunable tweezers can be obtained by tuning (controlling) the add (control) port input signal, in which the required number of microscopic volume (atom/photon/molecule) can be obtained and seen at the drop/through ports, otherwise, they propagate within a PANDA ring before collapsing/decaying into the waveguide.

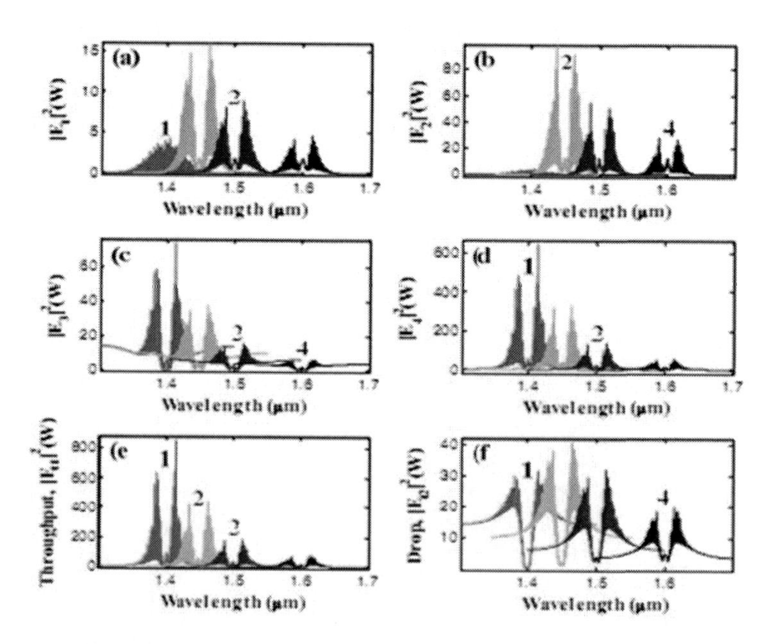

Figure 10.4. Simulation result of the tunable and amplified tweezers by varying the coupling coefficients.

In application, the trapped microscopic volumes can transport into the wavelength router via the through port, while the retrieved microscopic volumes are received via the drop port (connecting target), which can perform the drug delivery applications. The advantage of the proposed system is that the transmitter and receiver can be fabricated on-chip and alternatively operated by a single device. Result of the dynamic tweezers with microscopic volumes is as shown in Figure 10.5, where the generated wavelengths are 1.4, 1.45, 1.5, 1.55 and 1.6 μm, in which the manipulation of trapped microscopic volumes within the optical tweezers is as shown in Figure 10.5(d), where in this case study, the coupling coefficients are given as $\kappa_0 = 0.1$, $\kappa_1 = 0.35$, $\kappa_2 = 0.1$ and $\kappa_3 = 0.2$, respectively.

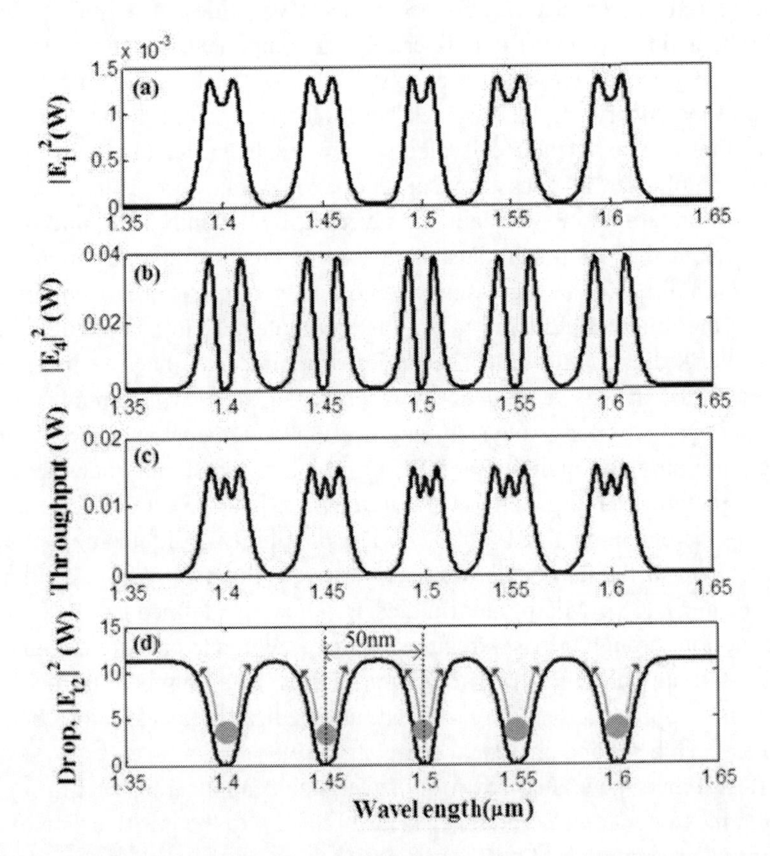

Figure 10.5. Result of the dynamic tweezers with microscopic volumes, where the generated wavelengths are 1.4, 1.45, 1.5, 1.55 and 1.6 μm.

10.4. MOLECULAR DIAGNOSIS

Alzheimer's disease (AD) is the most common form of dementia in the aging. The loss of neural cells associated with Alzheimer's disease is believed to be caused by the accumulation of beta amyloid plaques [24-25], tangles inside neuron [26] and genetic [24] are cause of abnormal axon transport in neuronal networks, which are important in the brain. Study of axonal transport began with the observation nerve cell bodies [27]. Axons are consisted of many microtubules. Each microtubule is a hollow cylindrical tube with outside diameter about 25 nm and a wall thickness of approximately 4 nm, When viewed in cross section, microtubules are seen to consist of 13 protofilaments with the fastest axonal transport occurring at velocities of 5 μm per second [28]. The main mechanism to deliver cellular components to their action site is long-range microtubule-based transport. The two major components of the transport machinery are the "engines," or molecular motors.

In the case of Alzheimer's disease new evidence is emerging that links proteins implicated in disease causation to axonal transport [29]. Microtubules in the axon are more resistant to severing by katanin than microtubules elsewhere in the neuron the sustained loss of tau from axonal microtubules over time renders them more sensitive to endogenous severing proteins, thus causing the microtubule array to gradually disintegrate in tauopathies such as Alzheimer's disease [30-32]. The optical trapping was the first invented by Ashkin [33]. It has emerged as a powerful tool with broad reaching applications in biology, physics, engineering and medicine. The ability of optical trapping and manipulation of viruses, living cell and bacteria without nondestructive and organelle without damage [34, 35], by laser radiation pressure were demonstrated [36, 37].. The possibility of liquids transportation and delivery at the nanoscale has seen rapid development within capillary or microchannel [38]. Micro-nanofluidics is generally defined as the study of fluid motion through or past structures with a size in one or more dimension in the 1-100 nm range [38]. Micro-nanofluidics is a burgeoning field with important applications in areas such as medical devices, biotechnology, chemical synthesis and analytical chemistry [39].

Researchers have studied in optical trapping application such as control kinesin movement on microtubule surface [40-42], to the positive side of axon transport. Erickson lab [43-47] have research interests revolving in the study of micro-nanofluidics and combination of optics are advancing flows, delivery and lead to implantable devices for living organs [48-50], where new advance in optics strategy using light is used to drive and halt neuronal activity with molecular specificity. The optical methods developed to date encompass a broad array of strategies, including photorelease of caged neurotransmitters, engineered light-gated receptors and channels and naturally light-sensitive ion channels pumps [51] and artificial neural networks [52].

Recently, the promising technique of microscopic volume trapping and transportation within an add/drop multiplexer have been reported both in theory [3] and experiment [53], respectively. Here the transporter is known as an optical tweezer. The optical tweezer generation technique is used as a powerful tool for manipulation of micrometer-sized particles. To date, the useful static tweezers are well recognized and realized. Moreover, the use of dynamic tweezers is now also realized in practical work [54-56]. Schulz *et al* [57] have shown the possibility of trapped atoms transfer between two optical potentials. In principle, an optical tweezer uses the forces exerted by intensity gradients in the strongly focused beams of light to trap and move the microscopic volumes of matter by a combination of forces induced by the interaction between photons, due to the photon scattering effects. In application, the field intensity can be adjusted and tuned, to the desired gradient field. The scattering force can then form the suitable trapping force. Hence, the appropriate force can be configured as the transmitter/receiver for performing a long distance microscopic transportation.

In operation, the optical tweezers can be trapped, transported and stored within the PANDA ring resonator and wavelength router, which can be used to form the microscopic volume (molecule) transportation, drug delivery via the waveguide [58], in which the manipulation of trapped microscopic volumes within the optical tweezers has been reported.

Optical trapping is one of the most powerful single-molecule techniques and application to living cell and important biological applications of tweezers is in the study of molecular motors [59] for useful in medicine. Chie Hosokawa et al [60] have demonstrated optical trapping of synaptic vesicles in a hippocampal neuron and found the intracellular synaptic vesicles can be trapped at the focal spot within the laser irradiation time. This occurs because the vesicles form clusters in a neuron, and these clusters are effectively trapped at the focal spot because of high polarizability.

In this section, we propose the optical trapping tools in the application of tangle protein in neuronal cell (motor protein, tau- protein, beta amyloid) which caused of Alzheimer's disease, Amyloid and tau protein are both implicated in memory impairment, mild cognitive impairment (MCI), and early Alzheimer's disease (AD), but whether and how they interact is unknown [61]. The optical tweezer can be trapped, transported, and stored within PANDA ring resonator [3] incorporating a wavelength router in the same drug delivery network system [30], we used the theory of optical trapping and transportation technique [62-64] for trap kinesins, plaques (tangle protein). Kinesin motor molecules are spheres and directly move onto microtubules where they could be activated by ATP [59].

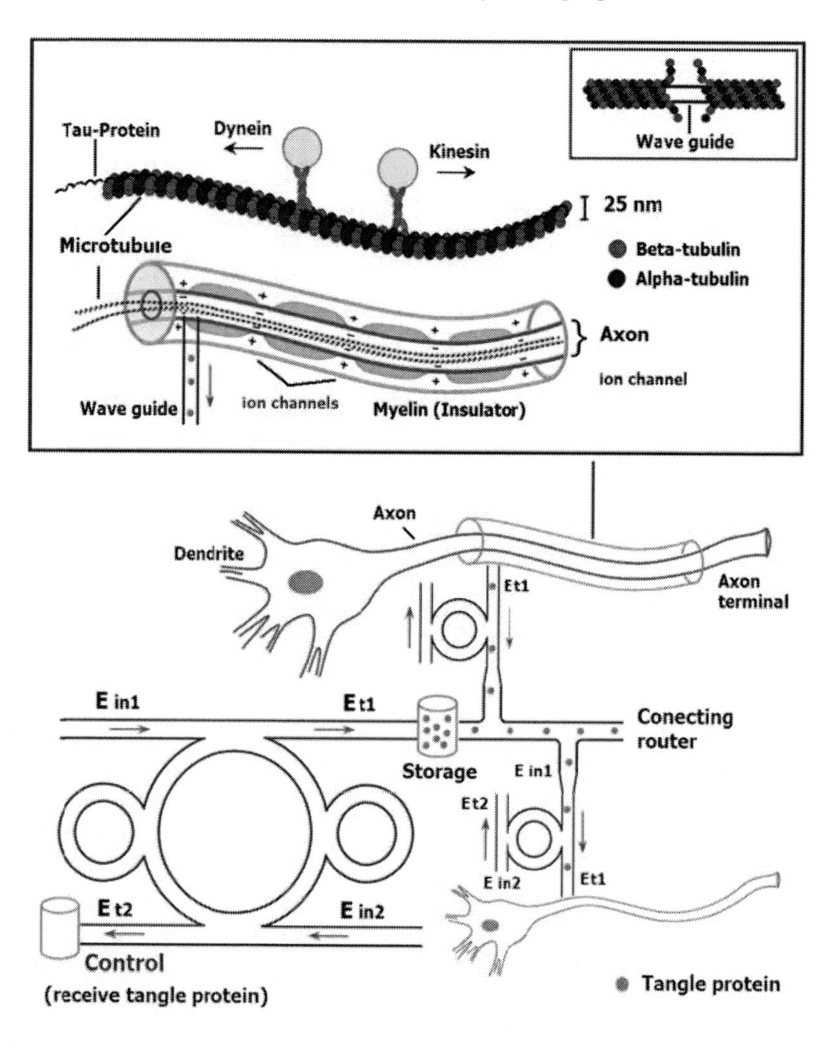

Figure 10.6. Schematic diagram of a Alzheimer's diagnosis system, where (a) a PANDA ring resonator, (b) a wavelength router, where R_{ad} is the add/drop filter radius.

We design the Alzheimer's diagnosis system in Figure 10.6, we put the end of throughput port (Et1) of add/drop single ring resonator and insert into microtubule, in which effective area of the waveguide is 1.13 x10^{-4} μm^2 (r = 6 nm) and the diameter of microtubule is 25 nm outside [28]. Figure 2 has shown schematic of inner waveguide inside microtubule, we design the system for the most useful tool to trap kinesin [65] and move/store tangle protein. We put the light waveguide into liquid medium for trap kinesin and also trap/store/receive tangle protein for protect ungrouping of them, which is caused of Alzehimer's disease of aging [66]. The optical tweezer can be induce the mechanical unfolding and refolding of a single protein molecule in the absence and the presence of molecular chaperones [67] and noninvasive neuronal of adult [68].

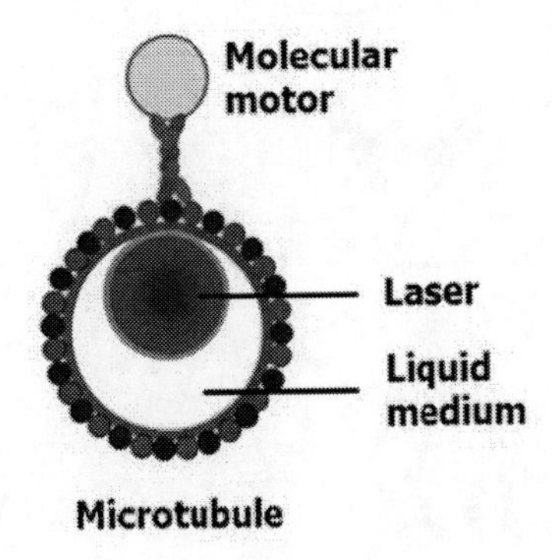

Figure 10.7. Cross section of microtubule and optical waveguide.

In simulation, the bright soliton with center wavelength at 1.50μm, peak power 2W, pulse 35fs is input into the system via the input port, the coupling coefficients are given as κ_0 = 0.5, κ_1 = 0.35, κ_2 = 0.1 and κ_3 = 0.35, respectively. The ring radii are R_{ad} = 20 and 15 μm, R_R = 5 and 6 μm and R_L = 5 and 6 μm, respectively. To date, the evidence of the practical device with the radius of 30 nm has been reported by the authors in reference [69] A_{eff} are 1.14 x 10^{-4} μm^2: In this case, the dynamic tweezers (gradient fields) can be in the forms of bright solitons, Gaussian pulses and dark solitons, which can be used to trap the required tangle protein, which is cause of Alzheimer's disease. There are four different center wavelengths of tweezers generated, where the dynamical movements are (a) $|E_1|^2$, (b) $|E_2|^2$, (c) $|E_3|^2$, (d) $|E_4|^2$, (e) through port and (f) drop port signals, where in this case all microscopic volumes are received by the drop port.

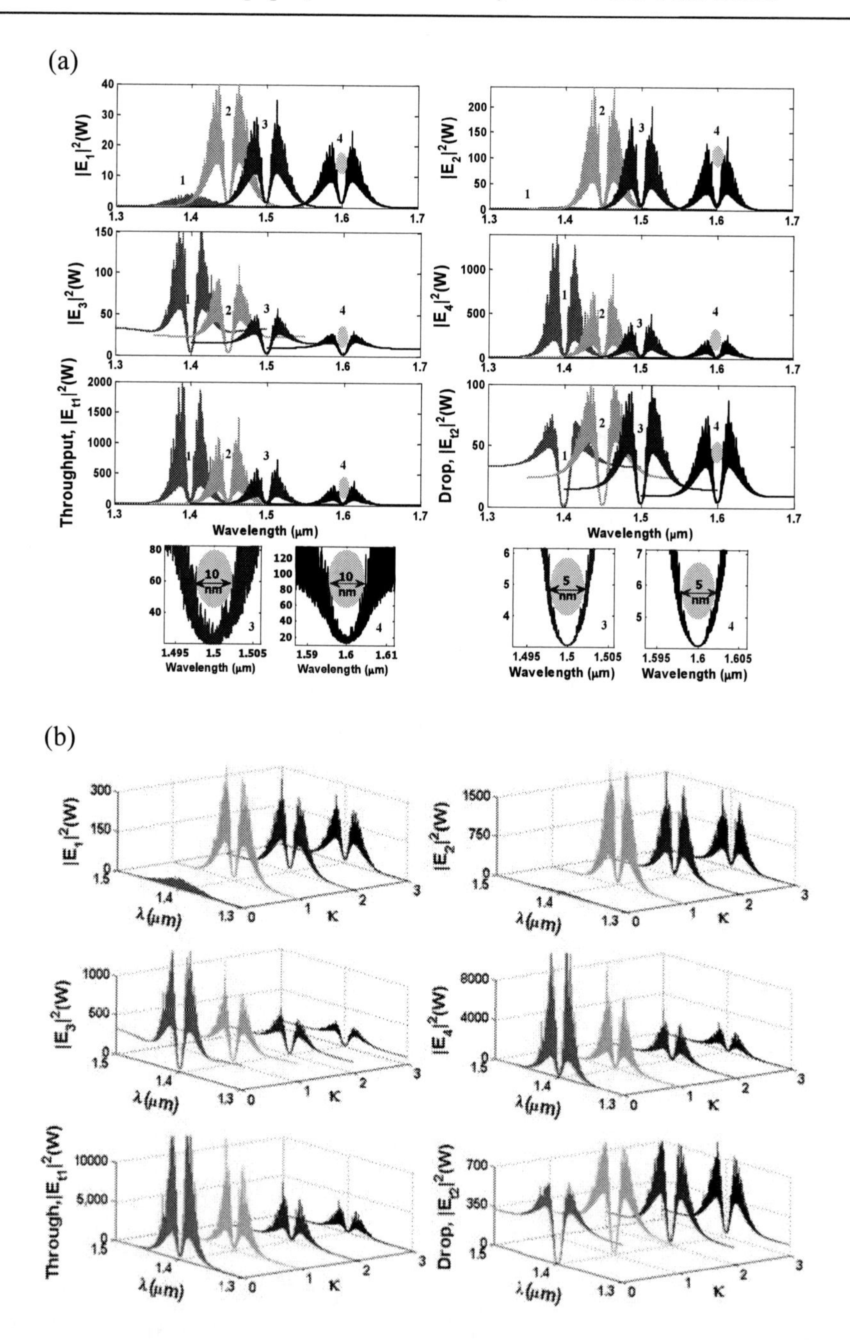

Figure 10.8. Result of the dynamic tweezers within the buffer with different (a) wavelengths and (b) coupling constants, where R_{ad} = 20 μm, R_R = R_L = 5 μm.

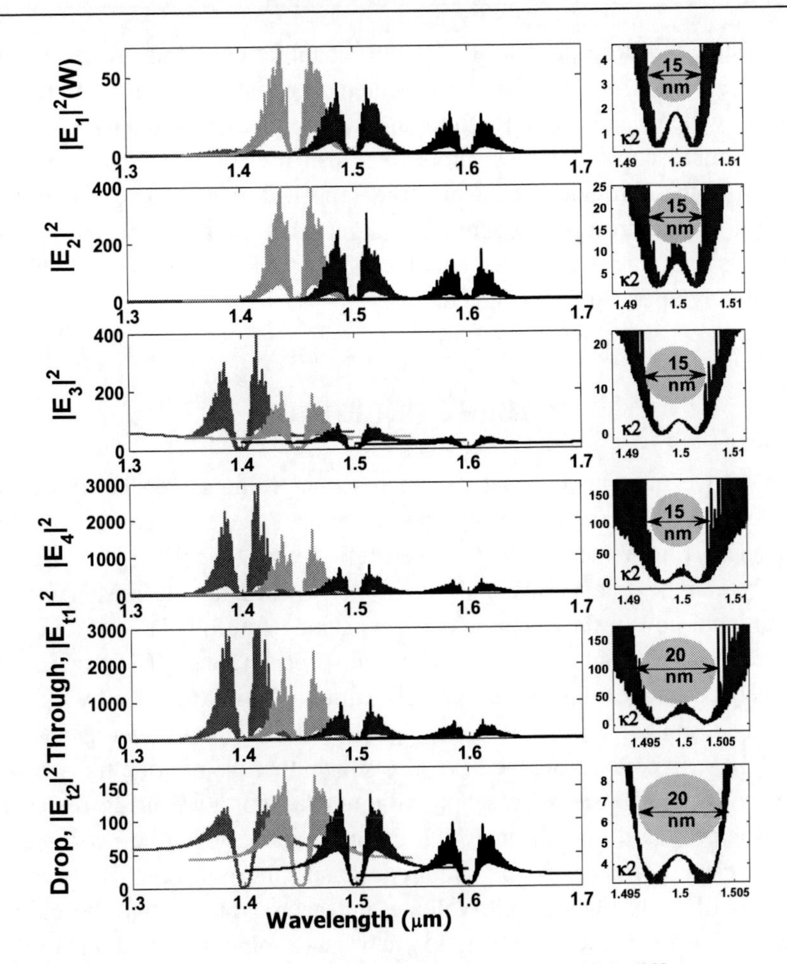

Figure 10.9. Result of the dynamic tweezers within the buffer with different wavelengths, where R_{ad} =15 μm, $R_R = R_L$ = 6 μm.

In practice, the fabrication parameters which can be easily controlled are the ring resonator radii instead of coupling constants. The important aspect of the result is that the tunable tweezers can be obtained by tuning (controlling) the add (control) port input signal, in which the required number of single protein (tau-protein/ beta myeloid, plaque)can be obtained and seen at the drop/through ports, otherwise, they propagate within a PANDA ring before collapsing/decaying into the waveguide.

More results of the optical tweezers generated within the PANDA ring are as shown in Figure 10.7, where in this case the bright soliton is used as the control port signal to obtain the tunable results. The output optical tweezers of the through and drop ports with different coupling constants are as shown in Figure 10.8(a), while the different wavelength results are as shown in Figure 10.8(b), which is allowed to form the selected targets. In application, the trapped microscopic volumes (molecules) can transport into the wavelength router via the through port, while the retrieved microscopic volumes are received via the drop port (connecting target). The advantage of the proposed system is that the transmitter and receiver can be fabricated on-chip and alternatively operated by a single device.

10.5. CONCLUSION

We have manipulated that the microscopic volumes can be trapped and transported into the optical waveguide by optical tweezers (vortices). By using a PANDA ring resonator and wavelength router, the long distance drug delivery can be transported and realized. By utilizing the reasonable dark soliton input power, the dynamic tweezers can be controlled and stored within the system. The obtained tweezer with free spectrum range (FSR) of 50 nm is achieved. Moreover, the tweezer amplification is also available by using the nanoring resonators and modulated signals via the control port as shown in Figure 10.9. In conclusion, we have also shown that the use of a transceiver to form the long distance microscopic volume transportation being realized by using the proposed system, in which the drug delivery can be performed via the wavelength router to the required (connecting) targets. However, the problems of large microscopic volume and neutral matter may be caused a problem, in which the searching for new guide pipe and medium [59], for instance, nano tube and specific gas will be the issue of investigation.

REFERENCES

[1] A. Rohrbach and Ernst H. K. Stelzer, " Trapping forces, force constants, and potential depths for dielectric spheres in the presence of spherical aberrations," *Appl. Opt.* 41, 2494-2507 (2002).

[2] K. Uomwech, K. Sarapat, and P.P. Yupapin, "Dynamic modulated gaussian pulse propatation within the double panda ring resonator system," *Microw. And Opt. Technol. Lett.* 52, 1818-1821 (2010).

[3] B. Piyatamrong, K. Kulsirirat, S. Mitatha, and P.P. Yupapin, "Dynamic potential well generation and control using double resonators incorporating in an add/drop filter," *Mod. Phys. Lett. B* 24(32): 3071–3082 (2010).

[4] H. Cai and A. Poon, "Optical manipulation and transport of microparticle on silicon nitride microring resonator –based add-drop devices," *Opt. Lett., Accepted*, August, 2010.

[5] A. Ashkin, J.M Dziedzic, J.E Bjorkholm, S. Chu, "Observation of a single-beam gradient force optical trap for dielectric particles," *Opt. Lett.* 11, 288-290(1986).

[6] K. Egashira, A. Terasaki, T. Kondow, " Photon-trap spectroscopy applied to molecules adsorbed on a solid surface: probing with a standing wave versus a propagating wave," *Appl. Opt.* 49, 1151-1157(2010).

[7] A.V. Kachynski, A.N. Kuzmin, H.E. Pudavar, D.S. Kaputa, A.N. Cartwright, P.N. Prasad, "Measurement of optical trapping forces by use of the two-photon-excited fluorescence of microspheres," *Opt. Lett.* 28, 2288-2290 (2003).

[8] M. Schulz, H. Crepaz, F. Schmidt-Kaler, J. Eschner, R. Blatt, "Transfer of trapped atoms between two optical tweezer potentials," *J. Modern Opt.* 54, 1619-1626 (2007).

[9] A. Ashkin, "Optical trapping and manipulation of neutral particles using lasers," *Proc. Natl. Acad. Sci.* 94, 4853–4860 (1997).

[10] Lindsey J.E. Anderson, E. Hansen, E.Y. Lukianova-Hleb, H.H Jason, D.O. Lapotko, J. Lindsey, "Optically guided controlled release from liposomes with tunable plasmonic nanobubbles," *J. Controlled Release* 144, 151-158 (2010).

[11] H. Shangguan, W. L. Casperson, A. Shearin, K.W. Gregory and S.A. Prahl, "Drug delivery with microsecond laser pulses into gelatin," *Appl. Opt.* 35, 3347-3358 (1996).

[12] M. Biondi, F. Ungaro, F. Quaglia, P. A. Netti, "Controlled drug delivery in tissue engineering," *Adv. Drug Deliv. Rev.* 60, 229–242(2008).

[13] Majeti N.V. Ravi Kumar,"Nano and Microparticles as Controlled Drug Delivery Devices," *J. Pharm. Pharmaceut. Sci.* 3(2), 234-258(2000).

[14] M.Z hang, T. Tarn, Ning Xi "Micro-/Nano-devices for Controlled Drug Delivery," in *Proceeding of the International conference on Robotics 6 Automation*, New Orleans. LA., (2004), pp.2068-2063.

[15] G. Huanga, J. Gaoa, Z. Hua, John V. St. Johnb, C.P. Bill, D. Morob,"Controlled drug release from hydrogel nanoparticle networks,"*J.Controlled Release* 94, 303-311 (2004).

[16] K. Svoboda and S.M. Block, "Biological applications of optical forces," *Annu. Rev. Biophys. Biomol. Struct.* 23, 247-283(1994).

[17] K. Dholakia, M. Gu,"Optical micromanipulation," *Chem.Soc.Rev.*, 37, 42-55 (2008).

[18] M. Tasakorn C. Teeka, R. Jomtarak, P. P. Yupapin, Multitweezers generation control within a nanoring resonator system, *Opt. Eng.*, 49, 075002(2010).

[19] S. Mitatha, N. Pornsuwancharoen, P.P. Yupapin, "A simultaneous short wave and millimeter wave generation using a soliton pulse within a nano-waveguide," *IEEE J. of Photon.Technol.Lett.* 21, 932-934 (2009).

[20] Y. Kokubun, Y. Hatakeyama,M. Ogata, S. Suzuki, N. Zaizen, "Fabrication technologies for vertically coupled microring resonator with multilevel crossing busline and ultracompact-ring radius," *IEEE J. of Sel. Top. Quantum Electron.* 11, 4-10(2005).

[21] P.P. Yupapin, W. Suwancharoen, "Chaotic signal generation and cancellation using a micro ring resonator incorporating an optical add/drop multiplexer," *Opt. Commun.* 280(2), 343-350(2007).

[22] P.P. Yupapin, P. Saeung, C. Li, "Characteristics of complementary ring-resonator add/drop filters modelling by using graphical approach," *Opt. Commun.* 272, 81-86(2007).

[23] J. Zhu, S.K. Ozdemir, Y.F. Xiao, L. Li, L. He, D.R. Chen, L. Yang, "On-chip single nanoparticle detection and sizing by mode splitting in an ultrahigh-Q microresonator," *Nature Photonics* 4, 46-49(2010).

[24] D.J. Selkoe, "Alzheimer's Disease: Genes, proteins, and therapy," *Phys. Rev.* 81(2):741-765 (2001).

[25] L. Jeffrey, J.L. Cummings, "Alzheimer's disease," *N. Engl. J. Med.* 351:56-67 (2004).

[26] R. Mayeux and M.D. Early "Alzheimer's disease," *N. Engl. J. Med.* 23:2194-2201 (2010).

[27] J.E. Morgan, "Circulation and axonal transport in the optic nerve," E*ye* 18:1089–1095 (2004)

[28] G. Karp, "*Cell and molecular biology*," John Wiley and Sons, Inc. 6th editon USA. (2010)

[29] G.B. Stokin, L.S.B. Goldstein, "Axonal transport and Alzheimer's disease," *Annu. Rev. Biochem.* 75:607-627 (2006).

[30] K.J. De Vos, A.J. Grierson, S. Ackerley, and C.C.J. Miller, "Role of axonal transport in neurodegenerative diseases," *Annu. Rev. Neurosci.* 31:151–173 (2008).

[31] D.B. Flaherty, J.P. Soria, H.G. Tomasiewicz, and J.G. Wood, "Phosphorylation of human tau protein by microtubule-associated kinases: GSK3b and cdk5 are key participants," *J. Neuroscience Res.* 62:463–472 (2000).

[32] W. Yu and P.W. Baas, "Changes in microtubule number and length during axon differentiation," *J. Neuroscience* 14(5):2818-2829 (1994).

[33] A. Ashkin, J.M. Dziedzic, and T. Yamane, "Observation of a single-beam gradient force optical trap for dielectric," *Opt. Lett.* 11:288-290 (1986).

[34] S.J. Lu, F. Qiang, J.S. Park, L. Vida, B.S. Lee, M. Strausbauch, P.J. Wettstein, G.R. Honig, and R. Lanza, "Biological properties and enucleation of red blood cells from human embryonic stem cells," *Blood* 112:4362-4363 (2008)

[35] H.D. Chen, K. Ge, Y. Li, J.J. Wu, Y. Gu, H.H. Wei, and Z. Tian, "Application of optical tweezers in the rresearch of molecular interaction between lymphocyte function associated antigen-1 and its monoclonal antibody," *Cellular and Molecular Immunology* 4:221-225 (2007).

[36] A. Ashkin and J.M. Dziedzic, "Optical trapping and manipulation of viruses and bacteria," *Science* 235:1517-1520 (1987).

[37] X. Zhao, Y. Sun, J. Bu, S. Zhu, and X.C. Yuan, "Microlens array enabled on-chip optical trapping and sorting," *Applied Opt.* 50(3):318-322 (2011).

[38] R.V. Harrison, N. Harel, J. Panesar, and R.J. Mount, "Blood capillary distribution correlates with hemodynamic-based functional imaging in cerebral cortex," *Cerebral Cortex* 12: 225-233 (2002).

[39] D. Psaltis, S.R. Quake, and C. Yang, "Developing optofluidic technology through the fusion of microfluidics and optics," *Nature* 442(27):381-386 (2006).

[40] M.J. Schnitzer and S.M. Block, "Kinesin hydrolyses one ATP per 8-nm step," *Nature* 388:386-390 (1997).

[41] S.M. Block, C.L. Asbury, J.W. Shaevitz, and M.J. Lang, "Probing the kinesin reaction cycle with a 2D optical force clamp," *PNAS* 100(5):2351–2356 (2003).

[42] M. Block, "Kinesin Motor Mechanics: binding, stepping, tracking, gating, and limping," *Biophys. J.* 92:2986–2995 (2007).

[43] A.J.H. Yang and D. Erickson, "Optofluidic ring resonator switch for optical particle transport," *Lab. Chip*. 10:769-774 (2010).

[44] A.J. Chung, Y.S. Huh, and D. Erickson, "A robust, electrochemically driven microwell drug delivery system for controlled vasopressin release," *Biomed. Microdevices* 11:861-867 (2009).

[45] M. Krishnan, M. Tolley, H. Lipson, and D. Erickson, "Hydrodynamically tunable affinities for fluidic assembly," *Langmuir* 25:3769-3744 (2009).

[46] A.J. Chung and D. Erickson, "Engineering insect flight metabolics using immature stage implanted microfluidics," *Lab. Chip*. 9: 669-676 (2009).

[47] B.S. Schmidt, A.H.J. Yang, D. Erickson, and M. Lipson, "Optofluidic trapping and transport on solid core waveguides within a microfluidic device," *Optics Express* 1(22):14322-14334 (2007).

[48] M. Segev, D.N. Christodoulides, and C. Rotschild, "*Method and system for manipulating fluid medium,*" US 2011/0023973 A. 2011 Feb 3.

[49] M. Bugge and G. Palmers, "*Implantable device for utiliztion of the hydraulic energy of the heart,*" US RE41,394 E. 2010 Jue 22.

[50] S.Y. Chen, S.H. Hu, D.M. Liu, and K.T. Kuo, "*Drug delivery nanodevice, its preparation method and used there of,*" US 2011/0014296 A1. 2011 Jan 20.

[51] S. Szobota and E.Y. Isacoff, "Optical control of neuronal activity," *Annu. Rev. Biophys*. 39:329–348 (2010).

[52] N. Suwanpayak, M.A. Jalil, C. Teeka, J. Ali, and P.P. Yupapin, "Optical vortices generated by a PANDA ring resonator for drug trapping and delivery applications," *Bio. Med. Opt. Express* 2(1): 159-168 (2011).

[53] H. Cai and A. Poon, "Optical manipulation and transport of microparticle on silicon nitride microring resonator –based add-drop devices," *Opt. Lett*. 35: 2855-2857 (2010).

[54] A. Ashkin, J.M. Dziedzic, and T. Yamane, "Optical trapping and manipulation of single cells using infrared laser beams," *Nature* 330:769–771 (1987).

[55] K. Egashira, A. Terasaki, and T. Kondow, "Photon-trap spectroscopy applied to molecules adsorbed on a solid surface: probing with a standing wave versus a propagating wave," *App. Opt.* 80: 5113-5115 (1998).

[56] A.V. Kachynski, A.N. Kuzmin, H.E. Pudavar, D.S. Kaputa, A.N. Cartwright, and P.N. Prasad, "Measurement of optical trapping forces by use of the two-photon-excited fluorescence of microspheres," *Opt. Lett.* 28: 2288-2290 (2003).

[57] M. Schulz, H. Crepaz, F. Schmidt-Kaler, J. Eschner, and R. Blatt, "Transfer of trapped atoms between two optical tweezer potentials," *J. of Mod. Opt.* 54: 1619-1626 (2007).

[58] N. Suwanpayak and P.P. Yupapin, "Molecular buffer using a PANDA ring resonator for drug delivery use," *Int. J. Nanomed.* 6:575-580 (2011).

[59] A. Ashkin, "Optical trapping and manipulation of neutral particles using lasers," *Proc. Natl. Acad. Sci.* 94: 4853-4858 (1997).

[60] C. Hosokawa, S.N. Kudoh, A. Kiyohara, and T. Taguchi, "Optical trapping of synaptic vesicles in neurons," *Appl. Phys. Lett.* 98:163705-3.(2011).

[61] O.A. Shipton, J.R. Leitz, J. Dworzak, C.E.J. Acton, E.M. Tunbridge, D. Franziska, H.N. Dawson, M.P. Vitek, R. Wade-Martins, O. Paulsen, and M. Vargas-Caballero, "Tau protein is required for amyloid induced impairment of hippocampal long-term potentiation," *The Journal of Neuroscience* 31(5):1688 –1692 (2011).

[62] K. Svoboda, C.F. Schmidt, B.J. Schnapp, and S.M. Block, "Direct observation of kinesin stepping by optical trapping interferometry," *Nature* 365:721-727 (1993).

[63] M.B. Steven, S.B. Lawrence, J.S. Bruce, "Head movement by single kinesin molecules studied with optical tweezers," *Nature* 348:348 – 352 (1990).

[64] A.B. Kolomeisky, E. Michael, and M.E. Fisher, "Molecular Motors: A theorist's perspective," *Annu. Rev. Phys. Chem.* 58:675–695 (2007).

[65] P. Bechtluft, R.G.H. van Leeuwen, M. Tyreman, D. Tomkiewicz, N. Nouwen, H.L. Tepper, A.J.M. Driessen, and S.J. Tans, "Direct observation of chaperone-induced changes in a protein folding pathway," *Science* 318:1458-1461 (2007).

[66] H.D. Ou-Yand and M.T. Wei, "Complex fluids: Probing mechanical properties of biological system with optical tweezers," *Annu. Rev. Phys. Chem.* 61:421-440 (2010).

[67] X. Xia, Z. Hu, and M. Marquez, "Physically bonded nanoparticle network: a novel drug delivery system," *J. Controlled Release* 103:21-30 (2005).

[68] J.H. Choi, M. Wolf, V. Toronov, W.U.C. Polzonetti, D. Hueber, L.P. Safonova, R. Gupta, A. Michalos, W. Mantulin, and E. Gratton, "Noninvasive determination of the optical properties of adult brain: near-infrared spectroscopy approach," *J. Biomed. Opt.* 9(1): 221–229 (2004).

[69] T.A. Nieminen, H.R. Dunlop, amd N.R. Heckenberg, "Calculation and optical measurement of laser trapping forces on non-spherical particles," *J. Quant. Spectr. Rad. Transf.* 70(4–6):627–37 (2001).

Chapter 11

OPTICAL CRYPTOGRAPHY

11.1. INTRODUCTION

A PANDA ring resonator type has been successfully used to investigate the dynamic behavior of dark-bright soliton collision within the modified add/drop filter [1-4]. All optical devices are increasingly becoming importance as integrated components for advanced optical technology applications and is widely used as optical sensor, signal processing and optical communication. The communication security segment has been recognized as a promising tool for information that necessitates the security and privacy requirements due the large demand of the world networks. Today, the security schemes such as quantum and optical techniques have been widely used in many applications such as sensors [5], computing [6], communication [7] signal processing [8], especially optical device in security communications [9-11]. Recently, an optical device known as a microring resonator in the form of an optical add/drop filter has been found in many applications [12-14]. The authors have shown that the transmitted signals can be suppressed with the chaotic signals. The required signals could be retrieved by the add/drop filter and the encryption-decryption method by using the proposed optical design system. However, the search for new devices and techniques still remains. In this paper, the authors have proposed the use of key suppression and recovery for optical cryptography using a PANDA ring resonator for high security communication. The required key can be suppressed (buried) by the noisy signals and the required signals can be secured and recovered (retrieved) by the specific designed optical device. The required data can be encrypted and decrypted by the optical encryption decryption keys respectively, in which both keys can be generated by using the suppressed optical keys (LIP signals). In application, such a proposed method can be used to form the secure communication either as digital or analog communications.

11.2. PROPOSED MODELING

The proposed system consists of three optical devices namely a PANDA ring resonator, add-drop filter and Mach-Zehnder interferometer as shown in Figs. 11.1, 11.2 and 11.3. The transmitter part of optical cryptography system consists of a PANDA ring resonator, two add-drop filter and one Mach-Zehnder interferometer. The receiver consists of two add-drop filter and one Mach-Zehnder interferometer.

An add/drop filter and the double microring resonators known as a PANDA ring resonator, is shown in Figure 11.1. To perform the dark-bright soliton conversion, the dark and bright solitons are first input into the add/drop optical filter system. The input optical field (E_{in}) and the control port optical field (E_{con}) of the bright and dark soliton pulses are given by [15]

$$E_{in}(t) = A\,\mathrm{sech}\left[\frac{T}{T_0}\right]\exp\left[\left(\frac{z}{2L_D}\right)-i\omega_0 t\right], \tag{11.1}$$

$$E_{con}(t) = A\tanh\left[\frac{T}{T_0}\right]\exp\left[\left(\frac{z}{2L_D}\right)-i\omega_0 t\right], \tag{11.2}$$

Here A and z are the optical field amplitude and propagation distance, respectively. T is a soliton pulse propagation time in a frame moving at the group velocity, $T=t-\beta_1 z$, where β_1 and β_2 are the coefficients of the linear and second-order terms of Taylor expansion of the propagation constant. $L_D=T_0^2/|\beta_2|$ is the dispersion length of the soliton pulse. T_0 in equation is a soliton pulse propagation time at initial input (or soliton pulse width), where t is the soliton phase shift time, and the frequency shift of the soliton is ω_0. This solution describes a pulse that keeps its temporal width invariance as it propagates, and thus is called a temporal soliton. When the soliton peak intensity $(|\beta_2/\Gamma T_0^2|)$ is given, then T_0 is known. For the soliton pulse in the microring device, a balance should be achieved between the dispersion length (L_D) and the nonlinear length ($L_{NL}=1/\Gamma\phi_{NL}$), where $\Gamma=n_2k_0$, is the length scale over which dispersive or nonlinear effects makes the beam become wider or narrower. For a soliton pulse, there is a balance between dispersion and nonlinear lengths, hence $L_D = L_{NL}$.

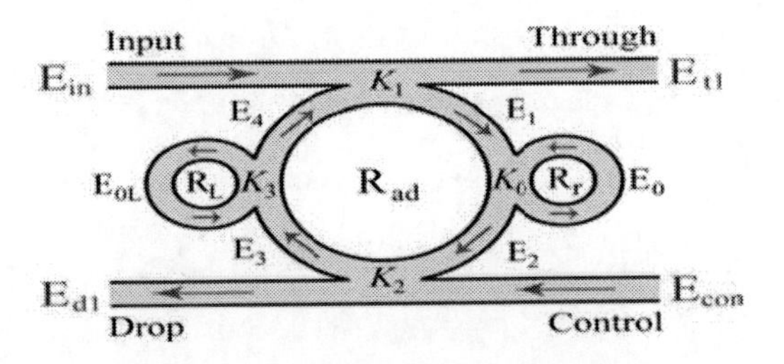

Figure 11.1. Schematic diagram of PANDA ring resonator.

In Figure 11.1, the PANDA ring resonator is used for the binary coded suppression. When the input light pulse passes through the first optical coupler of the add/drop optical multiplexing system, the transmitted and circulated optical fields can be written as [16]

$$E_{t1} = \sqrt{1-\gamma_1}\left[\sqrt{1-\kappa_1}E_{in} + j\sqrt{\kappa_1}E_4\right] \tag{11.3}$$

$$E_1 = \sqrt{1-\gamma_1}\left[\sqrt{1-\kappa_1}E_4 + j\sqrt{\kappa_1}E_{in}\right] \tag{11.4}$$

$$E_2 = E_0 E_1 e^{-\frac{\alpha}{2}\frac{L}{2} - jk_n\frac{L}{2}} . \tag{11.5}$$

Here κ_1 is the intensity coupling coefficient, γ_1 is the fractional coupler intensity loss, α is the attenuation coefficient, $k_n = 2\pi/\lambda$ is the wave propagation number, λ is the input wavelength light field and $L = 2\pi R_{ad}$, R_{ad} is the radius of add/drop device.

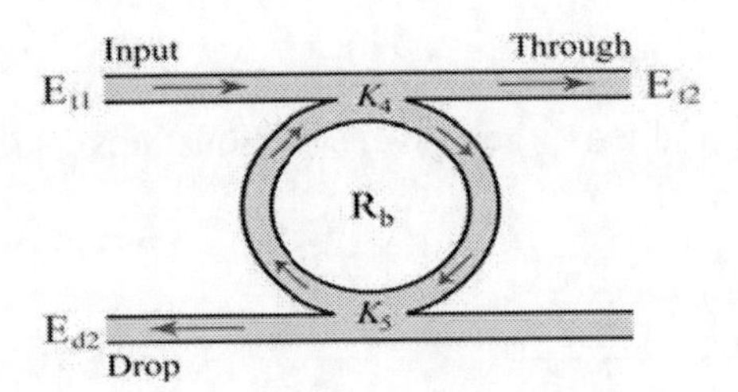

Figure 11.2. Schematic diagram of add/drop filter.

The optical fields at the second coupler of the add/drop optical multiplexing system are given by

$$E_{d1} = \sqrt{1-\gamma_2}\left[\sqrt{1-\kappa_2}E_{con} + j\sqrt{\kappa_2}E_2\right] \tag{11.6}$$

$$E_3 = \sqrt{1-\gamma_2}\left[\sqrt{1-\kappa_2}E_2 + j\sqrt{\kappa_2}E_{con}\right] \tag{11.7}$$

$$E_4 = E_{0L}E_3 e^{-\frac{\alpha}{2}\frac{L}{2} - jk_n\frac{L}{2}} \tag{11.8}$$

Here κ_2 is the intensity coupling coefficient, γ_2 is the fractional coupler intensity loss. The circulated light fields, E_0 and E_{0L} are the light field circulated components of the micro ring radii, R_r and R_L which couples into the right (RHS) and left sides (LHS) of the add/drop optical multiplexing system, respectively. The light field transmitted and circulated components in the right hand side (RHS) micro ring, R_r, are given by

$$E_2 = \sqrt{1-\gamma}\left[\sqrt{1-\kappa_0}E_1 + j\sqrt{\kappa_0}E_{r2}\right] \tag{11.9}$$

$$E_{r1} = \sqrt{1-\gamma}\left[\sqrt{1-\kappa_0}E_{r2} + j\sqrt{\kappa_0}E_1\right] \tag{11.10}$$

$$E_{r2} = E_{r1}e^{-\frac{\alpha}{2}L_1 - jk_nL_1} \tag{11.11}$$

Here κ_0 is the intensity coupling coefficient, γ is the fractional coupler intensity loss, α is the attenuation coefficient, $k_n = 2\pi/\lambda$ is the wave propagation number, λ is the input wavelength light field and $L_1 = 2\pi R_r$, R_r is the radius of right micro ring.

From Equations (11.9)-(11.11), the circulated roundtrip light fields of the (RHS) micro ring radii, R_r, are given in Equations (11.12) and (11.13), respectively

$$E_{r1} = \frac{j\sqrt{1-\gamma}\sqrt{\kappa_0}E_1}{1-\sqrt{1-\gamma}\sqrt{1-\kappa_0}e^{-\frac{\alpha}{2}L_1 - jk_nL_1}} \tag{11.12}$$

$$E_{r2} = \frac{j\sqrt{1-\gamma}\sqrt{\kappa_0}E_1 e^{-\frac{\alpha}{2}L_1 - jk_n L_1}}{1-\sqrt{1-\gamma}\sqrt{1-\kappa_0}e^{-\frac{\alpha}{2}L_1 - jk_n L_1}} \tag{11.13}$$

Thus, the output circulated light field, E_0, for the right micro ring is given by

$$E_0 = E_1 \left\{ \frac{\sqrt{(1-\gamma)(1-\kappa_0)} - (1-\gamma)e^{-\frac{\alpha}{2}L_1 - jk_n L_1}}{1-\sqrt{(1-\gamma)(1-\kappa_0)}e^{-\frac{\alpha}{2}L_1 - jk_n L_1}} \right\} \tag{11.14}$$

Similarly, the output circulated light field, E_{0L}, for the left hand side (LHS) micro ring at the left side of the add/drop optical multiplexing system is given by

$$E_{0L} = E_3 \left\{ \frac{\sqrt{(1-\gamma_3)(1-\kappa_3)} - (1-\gamma_3)e^{-\frac{\alpha}{2}L_2 - jk_n L_2}}{1-\sqrt{(1-\gamma_3)(1-\kappa_3)}e^{-\frac{\alpha}{2}L_2 - jk_n L_2}} \right\} \tag{11.15}$$

where κ_3 is the intensity coupling coefficient, γ_3 is the fractional coupler intensity loss, α is the attenuation coefficient, $k_n = 2\pi/\lambda$ is the wave propagation number, λ is the input wavelength light field and $L_2 = 2\pi R_L$, R_L is the radius of LHS micro ring.

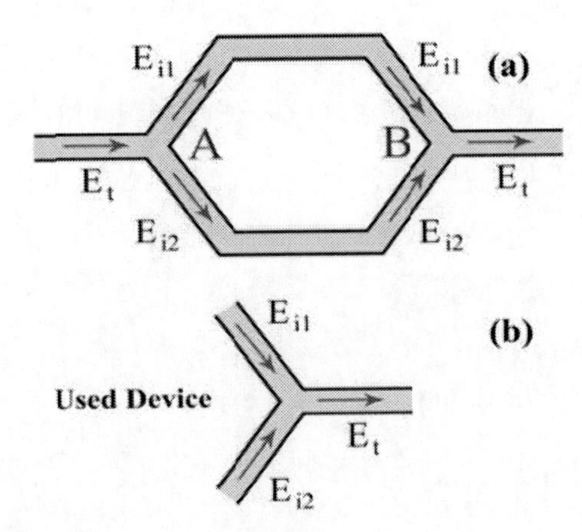

Figure 3. Schematic diagram of Mach-Zehnder interferometer.

From Equations (11.3)-(11.15), the circulated light fields, E_1, E_3 and E_4 are defined by given $x_1 = (1-\gamma_1)^{1/2}$, $x_2 = (1-\gamma_2)^{1/2}$, $y_1 = (1-\kappa_1)^{1/2}$, and $y_2 = (1-\kappa_2)^{1/2}$. Thus,

$$E_1 = \frac{jx_1\sqrt{\kappa_1}E_{in} + jx_1x_2y_1\sqrt{\kappa_2}E_{0L}E_{con}e^{-\frac{\alpha}{2}\frac{L}{2}-jk_n\frac{L}{2}}}{1 - x_1x_2y_1y_2E_0E_{0L}e^{-\frac{\alpha}{2}L-jk_nL}} \tag{11.16}$$

$$E_3 = x_2y_2E_0E_1e^{-\frac{\alpha}{2}\frac{L}{2}-jk_n\frac{L}{2}} + jx_2\sqrt{\kappa_2}E_{con} \tag{11.17}$$

$$E_4 = x_2y_2E_0E_{0L}E_1e^{-\frac{\alpha}{2}L-jk_nL} + jx_2\sqrt{\kappa_2}E_{0L}E_{con}e^{-\frac{\alpha}{2}\frac{L}{2}-jk_n\frac{L}{2}} \tag{11.18}$$

From Equations (11.3), (11.5), (11.16)-(11.18), the output optical field of the through port (E_{t1}) can expressed as

$$E_{t1} = x_1y_1E_{in} + \begin{pmatrix} jx_1x_2y_2\sqrt{\kappa_1}E_0E_{0L}E_1 \\ -x_1x_2\sqrt{\kappa_1\kappa_2}E_{0L}E_{i2} \end{pmatrix} e^{-\frac{\alpha}{2}\frac{L}{2}-jk_n\frac{L}{2}} \tag{11.19}$$

The power output of the through port (P_{t1}) is written by

$$P_{t1} = (E_{t1})\cdot(E_{t1})^* = |E_{t1}|^2. \tag{11.20}$$

Similarly, from Equations (11.5), (11.6), (11.16)-(11.18), the output optical field of the drop port (E_{d1}) is given by

$$E_{d1} = x_2y_2E_{con} + jx_2\sqrt{\kappa_2}E_0E_1e^{-\frac{\alpha}{2}\frac{L}{2}-jk_n\frac{L}{2}} \tag{11.21}$$

The power output of the drop port (P_{d1}) is expressed by

$$P_{d1} = (E_{d1})\cdot(E_{d1})^* = |E_{d1}|^2. \tag{11.22}$$

An add/drop optical filter device with the appropriate parameters is shown in Figure 2. The electric field detected by photodectector is given by [17]

$$E_{t2} = E_{t1}\frac{-\sqrt{1-\kappa_4}e^{-\frac{\alpha}{2}L_b - jk_n L_b} + \sqrt{1-\kappa_4}}{1-\sqrt{1-\kappa_4}\sqrt{1-\kappa_5}e^{-\frac{\alpha}{2}L_b - jk_n L_b}} \tag{11.23}$$

Here $L_b = 2\pi R_b$, R_b is radius of add/drop optical filter decoded as shown in Figure 2. The power output of the drop port (P_{t2}) is expressed by

$$P_{t2} = (E_{t2})\cdot(E_{t2})^* = |E_{t2}|^2 \tag{11.24}$$

The electric field detected by photodectector is given by

$$E_{d2} = E_{t1}\frac{-\sqrt{\kappa_4\kappa_5}e^{-\frac{\alpha}{2}\frac{L_b}{2} - jk_n\frac{L_b}{2}}}{1-\sqrt{1-\kappa_4}\sqrt{1-\kappa_5}e^{-\frac{\alpha}{2}L_b - jk_n L_b}} \tag{11.25}$$

The power output of the drop port (P_{d2}) is expressed by

$$P_{d2} = (E_{d2})\cdot(E_{d2})^* = |E_{d2}|^2 \tag{11.26}$$

The proposed system uses the Mach-Zehnder interferometer is as shown in Figure 11.3 The required optical cryptography is performed by incorporating the Mach-Zehnder interferometer. Considering the output (E_t) at point B which is equal to input 1 (E_{i1}) plus input 2 (E_{i2}), the electric field detected by a photodectector is given by [18]

$$E_t = E_{i1} + E_{i2}\ . \tag{11.27}$$

The power output of the drop port (P_{d2}) is expressed by

$$P_t = (E_t)\cdot(E_t)^* = |E_t|^2. \tag{11.28}$$

11.3. Key Suppression And Recovery

In simulation for optical key suppression, the used parameters of a PANDA ring resonator are fixed to be $\kappa_0 = 0.1$, $\kappa_1 = 0.2$, $\kappa_2 = 0.2$, and $\kappa_3 = 0.1$ respectively. The ring radii are $R_{ad} = 200\mu m$, $R_r = 15\mu m$, and $R_L = 15\mu m$. A_{eff} are 0.50, 0.25 and 0.25 μm^2 [19] for PANDA ring resonator, right and left micro ring resonators, respectively. For optical key recovery, the parameters of add/drop filter are fixed to be $\kappa_4 = 0.5$, $\kappa_5 = 0.2$, R = 100μm and $A_{eff} = 0.25$ μm^2, respectively. Moreover, our optical key suppression and recovery system should be possible to be fabricated, which can be confirmed by using the practical device parameters. Simulation results of the optical key signal with center wavelengths are at $\lambda_0 = 1.50\mu m$.

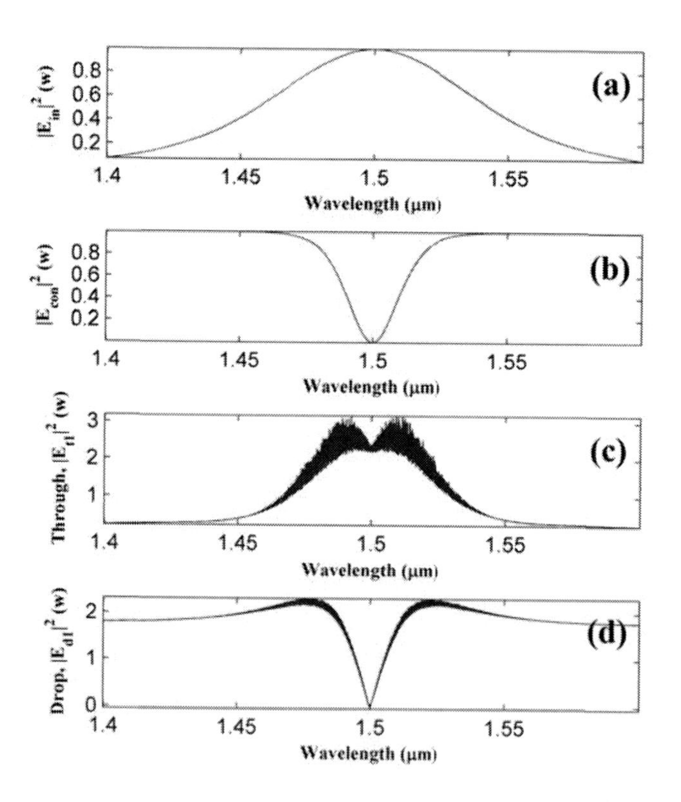

Figure 11.4. Simulation result for optical key suppression, where (a) $|E_{in}|^2$, (b) $|E_{con}|^2$, (c) $|E_{t1}|^2$ and (d) $|E_{d1}|^2$, where $R_r = 15\mu m$, $R_L = 15\mu m$, $R_{ad} = 200\mu m$ and $\alpha = 5 \times 10^{-5}$ $dBmm^{-1}$.

Figure 11.4 shows the simulation result for optical key suppression, where (a) $|E_{in}|^2$, (b) $|E_{con}|^2$, (c) $|E_{t1}|^2$ and (d) $|E_{d1}|^2$, where $R_r = 15\mu m$, $R_L = 15\mu m$, $R_{ad} = 200\mu m$ and $\alpha = 5 \times 10^{-5}\ dBmm^{-1}$. Here (a) is the input port for optical key suppression. The bright soliton pulse with 1W peak power is input into the input port. Figure 11.4(b) is the control port output that uses the dark soliton pulse with 1W peak power. The power output of the drop port is shown in Figure 11.4(d). Figure11.4(c) shows the power outputs of the through port which is the optical key suppression signal that transmitted to the receiver for secure optical communication and that can be formed as the reference signal in communication. Moreover, the peak power outputs of the through port and drop port are 2.3 and 3.2 W, respectively. They are larger than the input light pulse due to the optical nonlinear effects.

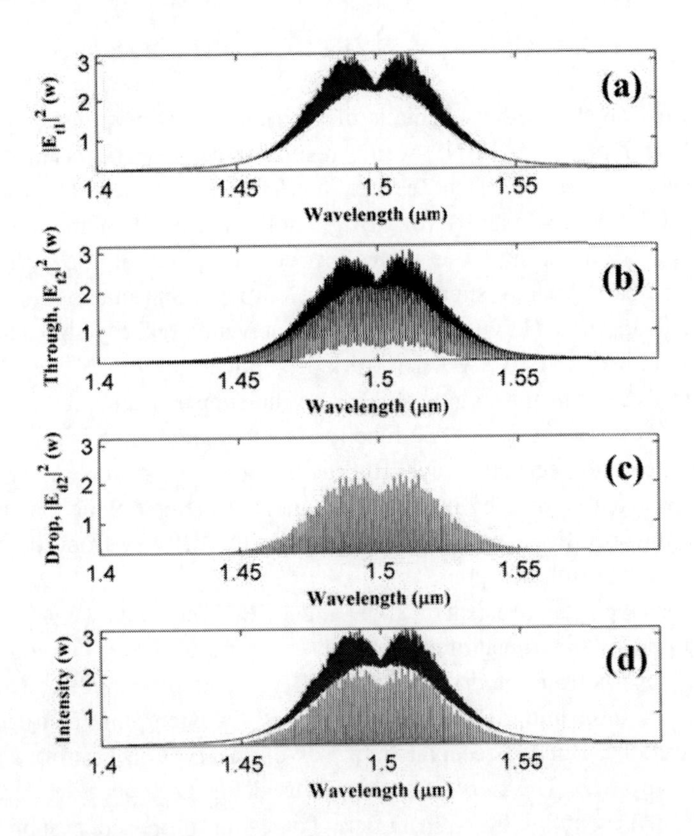

Figure 11.5. Simulation result for optical key recovery, where (a) $|E_{t1}|^2$, (b) $|E_{t2}|^2$, (c) $|E_{d2}|^2$ and (d) comparison of key suppression and key recovery, where $R_b = 100\mu m$ and $\alpha = 5 \times 10^{-5}$ dBmm^{-1}.

Figure 11.5 shows the simulation result for optical key recovery, where a) $|E_{t1}|^2$, (b) $|E_{t2}|^2$, (c) $|E_{d2}|^2$ and (d) comparison of suppression and recovery keys, where $R_b = 30\mu m$ and $\alpha = 5 \times 10^{-5}$ $dBmm^{-1}$. Here (a) is the input port for optical recovery key. The suppression key signal which looks like noisy signal is input into this port is actually the highly secure optical communication. Figure 11.5(b) and Figure 11.5(c) shows the power output of the through and drop ports for signal recovery, respectively. The power output from drop port is the analog signal which is sent by a sender is used as an optical key in the cryptography system referred to as the LIP signal (key). Figure 11.5(d) shows the comparison of suppressed optical key (blue line) and recovered optical key (red line), in which the secret signals are hidden by noisy signals.

11.4. Optical Cryptography System

Figure 11.6 shows the schematic diagram of optical cryptography system, where PA refers to the PANDA ring resonator device, AD is an Add/drop filter device and MZ is Mach-Zehnder interferometer device. The transmitter consists of 1 PA, 2 AD and 1 MZ. Bright soliton pulse (E_{in}) and dark soliton pulse (E_{con}) is input into the input and control ports of the system (key suppression part) for key suppression. The output signal obtained is E_{t1}. It is the optical suppressed key which is sent to receiver as security signal as shown in Figure 11.4(c). Firstly, AD performs the optical key or LIP key which is generated at the transmitter side (E_{d2}) from the suppressed signal as shown in Figure 11.5(c). Secondly, the AD function is generated by the LIP key to form the encrypted and decrypted keys. But the transmitter uses the encryption key from the encryption data by MZ. This means that the data which is encrypted is sent to the specific receiver, in which finally the LIP key from the recovery key becomes the ciphertext.

The receiver part consists of 2 AD and 1 MZ. The signal (E_{t1}) is sent into the input port by the transmitter as shown in Figure 11.5(a). The output signal (E_{d2}) that departs from the drop port is the LIP key as shown in Figure 11.5(c), is sent by the transmitter. The LIP key is used for encrypting and decrypting key generations. But the receiver part uses the decryption key from the data encryption by MZ. The decryption key is used for the ciphertext decryption, which is also sent by the transmitter. Thus, our proposed system can be claimed as a new and novel security technique using optical cryptography design, in which the secret data can be in the form of analog or digital data signals. Moreover, this triple security functions can be realized when the security can be formed by using the suppressed optical key, the optical key changing in every data frame and new optical cryptography technique.

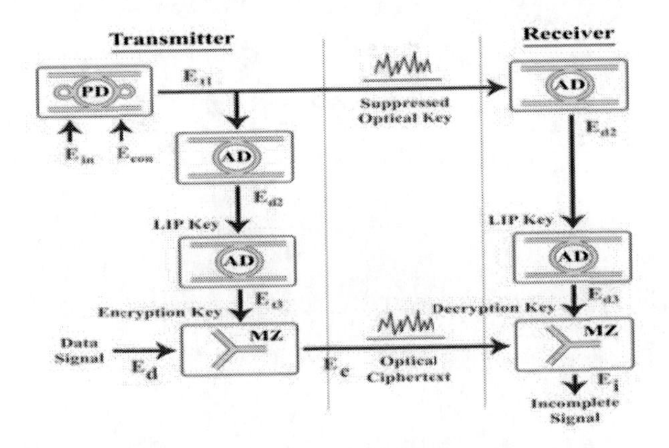

Figure 11.6. Schematic diagram of optical cryptography system, where PA: PANDA ring resonator device, AD: Add/drop filter device and MZ: Mach-Zehnder interferometer device.

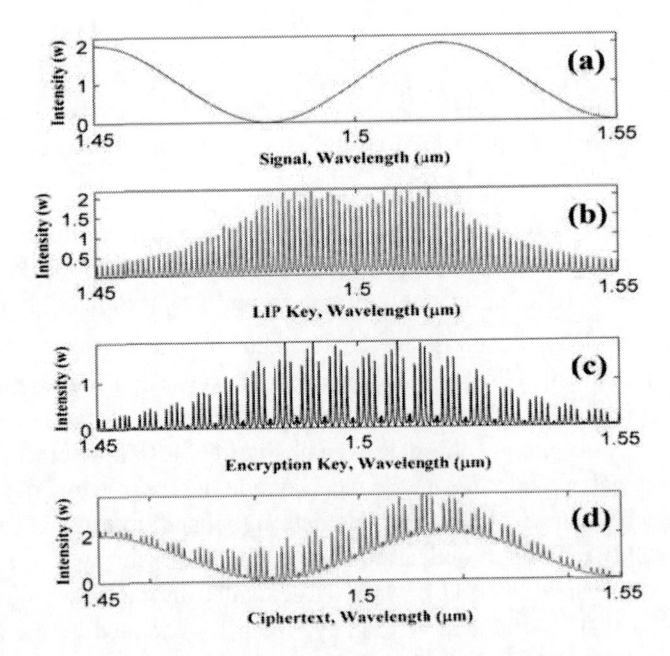

Figure 11.7. Simulation result of optical cryptography at sender side, where (a) data signal, (b) LIP signal(key), (c) encryption key and (d) ciphertext.

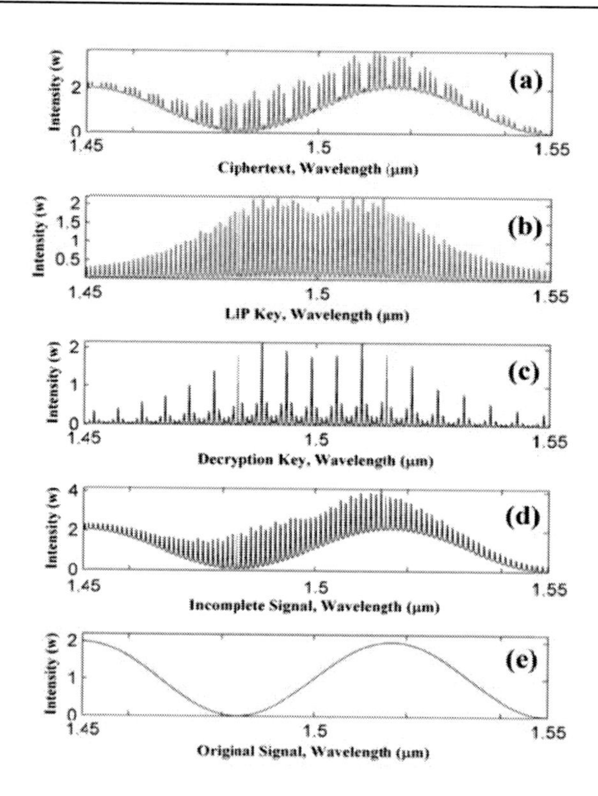

Figure 11.8. Simulation result of optical cryptography at receiver side, where (a) ciphertext, (b) LIP signal (key), (c) decryption key, (d) incomplete data signal and (e) original data signal.

Figure 11.7 shows the simulation result of optical cryptography at the sender side, where (a) data signal, (b) LIP key, (c) encryption key and (d) ciphertext. Here (a) is an example of data signal (E_d). Figure 11.7(b) shows the LIP key which is generated by the key recovery par. Figure 7(c) shows the encryption key which is generated by the LIP key using the add/drop filter device. Figure 11.7(d) shows the optical ciphertext which originates from the encrypted key to encrypt the data signal. Figure 11.8 shows the simulation result of optical cryptography at the receiver side, where (a) ciphertext, (b) LIP key, (c) decryption key, (d) incomplete data signal and (e) original data signal. Figure 11.8(a) shows the optical ciphertext which is sent by the transmitter. Figure 11.8(b) shows LIP key which is generated by the key recovery part. Figure 11.8(c) shows the decryption key which is generated by the LIP key (E_{t2}) using the add/drop filter device. Figure 8(d) shows the incomplete data (

E_i) signal which originates the decryption key to decrypt the ciphertext signal. Finally, the receiver can obtain the original signal (secret signal) as shown in Figure 11.8(e) by using the original data (E_d) = incomplete data (E_i) – LIP key (E_{t2}).

11.5. CONCLUSION

A new novel design for optical communication security by using the key suppression and recovery methods referred to as optical cryptography has been demonstrated by using the dark-bright soliton pair within a PANDA ring resonator. The proposed system can be fabricated using practical PANDA ring resonator parameters. The key suppression is designed based on a PANDA ring resonator in which the key recovery is obtained by using the add/drop filter. The optical cryptography is designed using the add/drop filter and Mach-Zehnder interferometer concept. The add/drop filter is used to generate the encryption, decryption keys from the LIP key and the Mach-Zehnder interferometer is used for signal combination. The secured communication functions of the proposed optical cryptography system can be realized by using the suppressed key, which can be changed (rearranged) in every data frame. In conclusion, the simulation results obtained have shown that the proposed system can indeed be achieved via the key suppression and recovery for optical cryptography system.

REFERENCES

[1] N. Suwanpayak, M. A. Jalil, C. Teeka, J. Ali, and P. P. Yupapin, " Optical vortices generated by a PANDA ring resonator for drug trapping and delivery applications," *Biomed. Opt. Express*, 2(1), 159-168(2011).

[2] P. P. Yupapin, N. Suwanpayak, B. Jukgoljun, and C. Teeka, "Hybrid transceiver using a PANDA ring resonator for Nanocommunication," *Physics Express*, 1(1), 1-9(2011).

[3] M. Tasakorn, C. Teeka, R. Jomtarak, and P. P. Yupapin, "Multitweezers generation control within a nanoring resonator system," *Opt. Eng.*, 49(7) 075002(2010).

[4] B. Jukgoljun, N. Suwanpayak, C. Teeka, and P.P. Yupapin, "Hybrid transceiver and repeater using a PANDA ring resonator for nano communication," *Opt. Eng.*, 49(12), 125003(2010).

[5] P. Hua, B.J. Luff, G. R. Quigley, J. S. Wilkinson, K. Kawaguchi, "Integrated optical dual Mach–Zehnder interferometer sensor", *Sensors and Actuators B*, 87, 250–257(2002).

[6] C. Kostrzewa, R. Moosburger, G. Fischbeck, B. Schuppert, K. Petermann, "Tunable polymer optical add/drop filter for multi-wavelength networks", *Photon. Technol. Lett.*, 9(11), 1487-1489(1997).

[7] P. D. Townsend, "Quantum cryptography on optical fiber networks", *Optical Fiber Technology*, 4(4), 345-370(1998).

[8] T. Carmon, T. J. Kippenberg, L. Yang, H. Rokhsari, S. Spillane, K. J. Vahala, "Feedback control of ultra-high-Q microcavities: Application to micro-Raman lasers and microparametric oscillators," *Opt. Express,* 13(9), 3558-3566 (2005).

[9] W. Siririth, S. Mitatha, O. Pingern, P.P. Yupapin, "A novel temporal dark-bright solitons conversion system via an add/drop filter for signal security use" *Optik- International Journal for Light and Electron Optics,* 121(21), 1955-1958(2010).

[10] B. Knobnob, S. Mitatha, K. Dejhan, S. Chaiyasoonthorn, P.P. Yupapin, "Dark–bright optical solitons conversion via an optical add/drop filter for signals and networks security applications", *Optik-International Journal for Light and Electron Optics*, 121(19), 1743-1747(2010).

[11] P.P. Yupapin, "Generalized quantum key distribution via micro ring resonator for mobile telephone networks", *Optik-International Journal for Light and Electron Optics*, 121(5), 422-425(2010).

[12] P. Gallion, F. Mendieta, S. Jiang, "Signal and quantum noise in optical communications and cryptography", *Progress in Opt.*, 52, 149-259(2009).

[13] Y. Dumeige, C. Arnaud, P. Féron, "Combining FDTD with coupled mode theories for bistability in micro-ring resonator", *Opt. Commun.,* 250(4-6), 376-383(2005).

[14] P. Rabiei, "Calculation of losses in micro-ring resonators with arbitrary refractive index or shape profile and its applications", *Lightw. Technol.,* 23(3), 1295-1301(2005).

[15] K. Sarapat, N. Sangwara, K. Srinuanjan, P.P. Yupapin and N. Pornsuwancharoen, "Novel dark-bright optical solitons conversion system and power amplification," *Opt. Eng.,* 48, 045004-1-7(2009).

[16] T. Phatharaworamet, C. Teeka, R. Jomtarak, S. Mitatha, and P. P. Yupapin, "Random binary code generation using dark-bright soliton conversion control within a PANDA ring resonator", *Lightw. Technol.*, 28(19), 2804–2809(2010).

[17] D. G. Rabus, M. Hamacher, U. Troppenz, and H. Heidrich, "Optical filters based on ring resonators with integrated semiconductor optical amplifiers In GaInAsP–InP", *IEEE J. Sel. Top. Quan. Elect.*, 8(6), 1405-1411(2002).

[18] A. Srivastava and S. Medhekar , "Switching of one beam by another in a Kerr type nonlinear Mach-Zehnder interferometer", *Opt. and La. Technol.*, 43(1), 29-35(2006).

[19] Y. Kokubun, Y. Hatakeyama, M. Ogata, S. Suzuki and N. Zaizen, "Fabrication technologies for vertically coupled microring resonator with multilevel crossing busline and ultracompact-ring radius," *IEEE J. Sel. Top. Quantum Electron.*, 11, 4-10(2005).

INDEX

D

E

F

G

H

I

L

M

N

O

P

Q

R

S

T

U

V

W